Refractions
of Mathematics
Education

Festschrift for Eva Jablonka

A volume in
Cognition, Equity, and Society: International Perspectives
Bharath Sriraman and Lyn English, *Series Editors*

International Perspectives on Mathematics Education—
Cognition, Equity, & Society

Bharath Sriraman and Lyn English
Series Editors

Refractions
of Mathematics
Education

Festschrift for Eva Jablonka

edited by

Christer Bergsten
Linköping University

Bharath Sriraman
University of Montana

≡IAP

INFORMATION AGE PUBLISHING, INC.
Charlotte, NC • www.infoagepub.com

Library of Congress Cataloging-in-Publication Data

Refractions of mathematics education : festschrift for Eva Jablonka / edited by Christer Bergsten, Linköping University, Bharath Sriraman, University of Montana.

    pages cm. – (Cognition, equity, and society : international perspectives)

  ISBN 978-1-68123-029-0 (pbk.) – ISBN 978-1-68123-030-6 (hardcover) – ISBN 978-1-68123-031-3 (ebook) 1. Mathematics–Study and teaching. 2. Mathematics–Philosophy. I. Jablonka, Eva (Mathematics education professor) II. Bergsten, Christer. III. Sriraman, Bharath.

  QA9.R344 2015

  510.71–dc23

                             2015008799

The cover photograph from 1990 by Eva Jablonka shows an interior part of the art house Tacheles, Berlin. Copyright © 2015 Eva Jablonka

CONTENTS

PREFACE

The conception of this book was born at the northern latitudes, during our discussions with a researcher who consistently pursued the ideas that came up, and put forward unexpected but deeply informed and relevant associations relocating the issues, as it were, within their proper theoretical landscapes. By her openness, humor, and expectancy of challenging and substantial responses from her co-discussants (also in such an informal setting), the exchange mode of our conversation produced advancements in our thinking around the issues we had discussed. Simplistic statements, whether naïve or covert, were doomed to succumb. The exclusive aim of the contributions was interrogation and clarification of the topical content at stake. For a focused engagement, however, the topic itself needed social relevance and matter.

The name of the researcher who so generously shared her knowledge in our discussions is Eva Jablonka. This *Festschrift* is dedicated to Eva Jablonka and her scholarly work in mathematics education; the academic spirit she represents, is becoming rarer in a field that is at risk of increased commodification and marketization. When reading her extensive and non-repetitive publications, be they on mathematical modeling, crosscultural classroom research and microanalysis of classroom discourse, curriculum theory and mathematical literacy, theorizing in mathematics education, or sociological analyses of mathematics education, one is also struck by the cohesiveness and development of her scholarship and the insights offered.

Commonly in academics, a *Festschrift* is a collection of papers that in some sense summarizes or acknowledges the work of a distinguished researcher in a specific field at a point of time when that person is, or is

Refractions of Mathematics Education, pages vii–viii
Copyright © 2015 by Information Age Publishing
All rights of reproduction in any form reserved.

close to, retiring. This *Festschrift* is different as Eva is far from retiring, and secondly, as the colleagues and friends who had been invited to contribute were free to choose the writing genre and topic, not necessarily related to her work, it was more important that they adhere to a spirit of intellectual integrity and scientific honesty, and a purpose worth pursuing for generating new knowledge. Because Eva is born Viennese, she also would not shred an ironic allusion. As editors, we wish to thank all the authors for their generous original contributions: a colorful panorama of refractions of mathematics education.

—**Christer Bergsten** and **Bharath Sriraman**

INTRODUCTION

Christer Bergsten and Bharath Sriraman

Als Hilfe, selbstständig den Blickwinkel zu ändern, und als Mittel zum Erkennen und Bewerten von Alternativen, zur Suche nach Gründen und beim Austausch von Argumenten ist Reflexion ... eine Haltung bei der Beurteilung mathematischer Methoden zur bewußten Gestaltung von Handlungssituationen, die dazu beiträgt, zu erkennen, daß man als Rezipient oder Betroffener autorisiert und als Anwender verpflichtet ist, eine kritische Haltung einzunehmen.

—Jablonka, 1996, p. 187

Although researchers have studied the conditions and outcomes of the teaching and learning of mathematics over a long time (Kilpatrick, 2014), the date of the first ICME congress in Lyon may be seen as the formal birthday of mathematics education as a research field *in its own right* (Furinghetti, Matos, & Menghini, 2013, p. 289). That date would place this book within a frame of reference of at least 45 years, an age one would hesitate to call mature for a scientific discipline. In the 1998 ICMI study on mathematics education as a research domain, the editors concluded that the "search for our common identity as researchers in mathematics education is not over" (Sierpinska & Kilpatrick, 1998, p. 547). From that perspective, the book *Didactics of Mathematics as a Scientific Discipline*, which was dedicated to Hans-Georg Steiner some years earlier, may be seen as a *state of the progress* rather than a description of the *state of the art* of this discipline at that time. In the preface the editors wrote that they did not only want

Refractions of Mathematics Education, pages ix–xiv
Copyright © 2015 by Information Age Publishing

to demonstrate the level reached and the maturity gained, but also to indicate questions that are still open and tasks that need to be solved in the future. Both Professor Steiner and the IDM would be honoured by showing that the object of their promotion is alive and well in both its international connections and its disciplinary diversions. (Biehler, Scholz, Sträßer, & Winkelmann, 1994, p. 6)

That description assumes an evolutionary notion of a maturing body of knowledge. Twenty years after the publication of those statements, is the 'body' still "alive and well in both its international connections and its disciplinary diversions?" Alive yes, as witnessed by the increasing bulk of texts potentially contributing to the formation of this body, as for example articles in international scientific journals whose names signify that they belong to this particular field. A simple count of all issues from 2013 of the 18 periodicals specializing in mathematics education (that were ranked based on a given set of criteria; see Törner & Arzarello, 2012) by European scholars in a survey collaboratively organized by the European Mathematical Society (EMS) and the European Society for Research in Mathematics Education (ERME), showed a total of about 500 articles published in 79 issues, adding up to some 10,000 pages. While this is still a relatively small number compared to other fields,[1] considering that it is only a fraction of all published work about the mathematics education research during this one year, there are a number of pertinent questions seldom asked that pop up. These include: who wrote these articles and why;[2] who printed these articles and why;[3] who read these articles and why; who cited these articles and why; and what contribution to the development and differentiation of this body of knowledge have these articles accomplished and for what purpose. Concurrent with these types of questions, Paul Ernest has suggested a postmodern view of research in mathematics education that, "does not separate research and knowledge from the group of people that do this research and produce this knowledge, and from the goals they attempt to reach through these" (Ernest, 1998, p. 80). There are also many other questions one should ask about what, in light of the questions above, in Foucaultian terms could be called the discursive formation of mathematics education; what regularity with regard to "order, correlations, positions and functionings, transformations" (Foucault, 2002, p. 41) can be described between its chosen themes and objects of study, the concepts it has developed, and the types of statements it has made? Only answers to questions like these would indicate the extent to which this 'body' is (still) well.

The term "diversions" in the quote from Biehler et al. (1994) seems to refer to

the differentiation of the theoretical framework of the didactics of mathematics, the diversification of methods used, and of the objects of interest in the international discussion. (Biehler et al., 1994, p. 6)

When manifested, rather, through adherence to different intellectual roots and theoretical orientations (Sriraman & English, 2010), 'diversions' may better be termed *refractions of mathematics education*. For example, the collection and analysis of empirical data in an educational study are by necessity, *refracted* through the specific analytical *lens* or approach chosen, as well as the aim of the study itself. Refractions can also refer to ways of looking at old phenomena through new lenses. Such refractions have the potential to generate what Jeremy Kilpatrick, when discussing the reasonable ineffectiveness of research in mathematics education, called landmark studies:

> A landmark research study is one that confronts us with data analyzed and organized so as to shake our preconceptions and force us to consider new conceptions. (Kilpatrick, 1981, p. 27)

This would produce new (theoretical) lenses as bases for renewed refractions in an iterative process, resonating with the following description by Alan Bishop of the purpose of research in mathematics education:

> Theory is the essential product of the research activity and theorizing is, therefore, its essential goal. (Bishop, 1992, p. 711)

Theorizing is a common core across Eva Jablonka's wide-ranging scholarship, both in empirical studies and philosophical writings. However, theorizing alone does not guarantee the employment of a critical stance toward the research, something for which one is not only authorized as a researcher, but should see as an obligation (see Jablonka's quote at the beginning, from her dissertation about mathematical models).

This *Festschrift* is a tribute to Jablonka and her work, which has provided refractions of mathematics education through the mathematical, the philosophical, the political, and the social dimensions generally treated as inseparable; most of them have been pursued by the authors of the chapters in this book. The discussion of issues in mathematics education as developed in the chapters by Peter Appelbaum on the elusive experience of encountering mathematics, Paul Dowling on the purification of theories, Michael Otte on the relationship between philosophy and mathematics, and Ole Skovsmose on ethnomathematics as discourse, may be seen as refracted through the *philosophical*, in a wide sense, while different aspects of the *political* frame the chapters by Uwe Gellert on messages from textbook images, Brian Greer on the politics of mathematics education, and Paola Valero and Alexandre Pais on the presence of the political in mathematics education research. The *mathematical* provides the lens for Christer Bergsten, Michael Fried, and Carl Winslow in their respective chapters on meaning and representations, mathematical models, and praxeologies for the calculus, thus constituting examples of mathematical approaches in mathematics education (Bergsten, 2014). Some aspects of research in mathematics education are refracted through the *personal* in Stephen Lerman's chapter, which is a conversation with Eva Jablonka. The final chapter, The White Chapter, will remain just a fragment of a refraction until readers start scribbling over the blank pages; the only rule being to include the existing picture, glossary, questionnaire, and references.

The four refractions (philosophical, political, mathematical, and personal) can be further viewed through a *Foucaultian* lens to discuss identity in mathematics education. One's experience as a researcher in mathematics education is subjective in a Foucaultian sense: both in engaging with the field, as well as the choices one makes in theorizing within one's work. As Foucault (1982) clarified the word *subject* carries a double meaning: one meaning that is characterized by dependence or control by someone else, and another meaning characterized by one's identity or self-knowledge. But power relations are evident in both meanings: "power which subjugates or makes subject to" (p. 212). In mathematics education, an entry-level researcher is often under the control of a senior professor in terms of the research agenda that they adhere to, particularly in large-scale studies that require funding from multinational sources. Power relations further crop up when a researcher encounters gatekeepers (editors, referees, etc.) who impose subjectivity in the publication process. For instance, adherence to an unfamiliar methodology or the employment of a theoretical lens that is not in vogue (e.g., constructivism in the 1990's), results in the rejection of one's research by the discipline.

A different pitfall that one encounters at a later stage as a researcher, is that of internal theorizing:

> There is a danger of 'internal' theorising without taking notice of the profound and ongoing development over long periods in psychology, philosophy, cultural psychology and anthropology, social linguistics and semiotics, and sociology. (Jablonka & Bergsten, 2010, p. 116)

Jablonka's career as a mathematics education researcher has experienced the vicissitudes of Foucaultian power relations with the outcome of a researcher that has broken away from the "subjugates or makes subject to" cycle of oppression.

NOTES

1. See Sriraman (2012) where also an analysis of the economic values involved is offered.
2. In a study of publishing practices of the UK mathematics education research community, using social network analysis Craig (2012) found interesting pat-

terns of author collaboration in journals, suggesting positioning strategies
within the field.

3. 51% of the articles counted above were published in journals owned by one
publisher and about 21% in journals owned by one other publisher.

REFERENCES

Bergsten, C. (2014). Mathematical approaches. In S. Lerman (Ed.), *Encyclopedia of
mathematics education* (pp. 376–383). New York, NY: SpringerReference.

Biehler, R., Scholz, R. W., Sträßer, R., & Winkelmann, B. (Eds.) (1994). *Didactics of
mathematics as a scientific discipline.* Dordrecht, the Netherlands: Kluwer.

Bishop, A. (1992). International perspectives on research in mathematics education. In D. A. Grouws (Ed.), *Handbook of research on mathematics teaching and
learning* (pp. 710–723). New York, NY: Macmillan.

Craig, A. J. (2012). *Publishing practices and the role of publication in the work of academics
in the mathematics education research community in England.* (Unpublished doctoral thesis) London, England: Institute of Education, University of London.

Ernest, P. (1998). A postmodern perspective on research in mathematics education. In A. Sierpinska & J. Kilpatrick (Eds.), *Mathematics education as a research
domain: A search for identity. An ICMI Study* (pp. 71–85). Dordrecht, the Netherlands: Kluwer.

Foucault, M. (1982). The subject and power. In H. Dreyfus & P. Rabinow. *Michel
Foucault: beyond structuralism and hermeneutics.* Chicago, IL: University of Chicago Press.

Foucault, M. (2002). *The archaeology of knowledge.* London, England: Routledge.

Furinghetti, F., Matos, J. M., & Menghini, M. (2013) From mathematics and education to mathematics education. In M. A. Clements, A. Bishop, C. Keitel, J. Kilpatrick, & F. Leung (Eds.), *Third international handbook of mathematics education*
(pp. 273–302). New York, NY: Springer.

Jablonka, E. (1996). *Meta-Analyse von Zugängen zur mathematischen Modellbildung und
Konsequenzen für den Unterricht.* [Meta-analysis of approaches to mathematical
modeling and consequences for teaching.] (Dissertation) Berlin, Germany:
Transparent Verlag H. & E. Preuß.

Jablonka, E., & Bergsten, C. (2010). Commentary on theories of mathematics education: Is plurality a problem? In B. Sriraman & L. English (Eds.), *Theories
of mathematics education: Seeking new frontiers* (pp. 111–117). New York, NY:
Springer.

Kilpatrick, J. (1981). The reasonable ineffectiveness of research in mathematics
education. *For the learning of mathematics, 2*(2), 22–29.

Kilpatrick, J. (2014). History of research in mathematics education. In S. Lerman
(Ed.), *Encyclopedia of mathematics education* (pp. 267–272). New York, NY:
SpringerReference.

Sierpinska, A., & Kilpatrick, J. (1998). Continuing the search. In A. Sierpinska &
J. Kilpatrick (Eds.), *Mathematics education as a research domain: A search for identity. An ICMI Study* (pp. 527–548). Dordrecht, the Netherlands: Kluwer.

Sriraman, B. (2012). Polarities in (Nordic) mathematics education: Scaling the field. In G. H. Gunnarsdóttir, F. Hreinsdóttir, G. Pálsdóttir, M. Hannula, M. Hannula-Sormunen, E. Jablonka, U. T. Jankvist, A. Ryve, P. Valero, & K. Wæge (Eds.), *Proceedings of the sixth nordic conference on mathematics education* (pp. 65–80). Reykjavík, Iceland: University of Iceland Press.

Sriraman, B., & English, L. (Eds.) (2010). *Theories of mathematics education: Seeking new frontiers.* New York, NY: Springer.

Törner, G., & Arzarello, F. (2012, December). Grading mathematics education research journals. *EMS Newsletter,* pp. 52–54.

Christer Bergsten
Linköping University
Sweden

Bharath Sriraman
University of Montana, Missoula
USA

CHAPTER 1

ON RETURNING

Peter Appelbaum

PART I

There is a kind of self-annihilation in mathematical work that makes it profoundly difficult to return to the world as oneself; and at the same time to
savor the mathematics with which one has been engaged. We had been
exploring star-like shapes that could be drawn without lifting a pencil off
the paper. We were working systematically, placing a number of points in a
circle, starting at one point, and then drawing lines to points a fixed number away around the circle. With seven points connected one by one, we
could produce a "routine" septagon; connecting two away led to a lovely
star; connecting three away, an even pointier star; four away felt like the
three-away star, but emerged "backwards"; five-away was "backwards" from
the two-away pattern; six-away was a backwards one-away. In that moment,
when fifth grader Glen twisted around and exclaimed, "You have to be at
least two away to get a star, and not just a shape, and it's gotta be possible
that the two-away dot isn't exactly half-way, so you don't get an asterisk, so,
so . . . so, . . . you gotta have a shape bigger than a square to get a star!" And
the entire class was looking at him, and listening, and understanding what
he meant. We also had already returned to the world, were out of that moment, and back in the classroom where we could, of course, think about
the implications of Glen's conjecture, or capture it in writing, or continue

Refractions of Mathematics Education, pages 1–13
Copyright © 2015 by Information Age Publishing

to collect examples of stars with different numbers of vertices. And we were already losing that same moment, losing the non-self experience of mathematics; we were returning both to each other, out of the mathematics, and back into the world of tasks and objectives.

There is a kind of scary vertigo in mathematical work that makes it profoundly difficult to allow mathematics to share one's experience of the world, to return to a time and place where one is no longer a self in relation to a mathematical idea or object, but rather is present with the mathematics. We had been performing small-group tableaux about infinity and zero in a university seminar. We were using sequences of frozen theater scenes that visually communicated each group's perspective: infinity and zero are synonymous, zero is more important than infinity, infinity is more important than zero, it all boils down to faith, or it all boils down to physical measurement. As each group performed, others were fascinated; they were confronting scenes nothing like those that they had moments before felt obvious and universally shared. At the same time, I (as the seminar leader) panicked; already in my shared excitement, I was also bewildered about how to sustain the engagement. The first tableau had been followed by silence. A request for comments went unanswered. I asked if what I thought about as I viewed the tableau, was anything like what the group had hoped for their audience. Yes. How amazing, that they could so excellently communicate such a complex set of ideas! Did anyone in the class think of something else? More silence. After the second tableau, I wondered if I should say anything or ask anything? Could an audience of tableau artists respond to each others' performances? Was this a matter of using teaching strategies, or should we embrace the difficulties of returning after the moment of complete engagement?

A common interpretation of such classroom events has been that the students are not identifying a context as mathematical. In this ordinary view, learners feel *more free* when they do not feel or see mathematics; when asked to represent mathematical ideas, they can do this well when they think what they are doing is not mathematical. Yet naming an activity as mathematical, is associated with an instant blockage that re-instantiates previous, negative relationships with mathematics. In this essay, I will ask us to consider what would happens if we did not assume mathematics was associated with the disruption of engagement. Instead, I want to explore the many ways in which mathematics classrooms undermine engagement by misunderstanding the challenges of returning from genuinely, generative mathematical encounters. A painful return, exacerbated by a well-meaning teacher yearning to transform an encounter into a representation of that encounter, if it is repeated over and over again in school contexts, it can cause students to develop well-crafted strategies of resistance and deflection; they will not encounter the pleasures of self-annihilation that are found in the engagement with mathematics, but will have the excruciating experience of return that follows such engagement.

The melancholic moral of mathematics education is that even seemingly successful pedagogy might work against its own goals; this could lead over time to the learner's increased resistance to mathematical thinking. This may seem counter to much of educational psychology, which imagines satisfaction and accomplishment as an intrinsic motivator. In Maslow's hierarchy of needs, self-actualization is assumed to be self-perpetuating. The issue with returning is that the rich experience of self-annihilation and mathematical relation is immediately followed, as if in a sadistic Pavlovian experiment, by the punishment of loss. Of course, if the joy and pleasure of the experience is powerful enough and intense enough, it could lead people to return to the mathematics over and over again, despite the pain of return out of the union; but this raises the question of where to draw the line between productive pleasure and addictive desire.

We have a responsibility when we wield the "weapon" of return from mathematical encounters, to take an ethical stance and do no harm: The return is an experience of splitting from the total union with mathematics and the loss of self. It is a mirror with which one is confronted once more with the recognition of oneself; it becomes an uncanny presentation of self as having just been "somewhere else," and now is back in the world where the self is once more a mere self who is distinct from others. But through the mirror now, one sees both oneself and someone else; a someone who is returning from another place. This uncanny return is at once disturbing and mesmerizing; vertiginous disorientation and an anxious anticipation (Appelbaum, 2011). The feeling is simultaneously one of fascination and fear; it is at once confusing and alluring. In a Freudian sense, the splitting from the union (and lack of self to separation) and the uncanny, is one of violence. Pedagogically we can see that the repeated encounter with the uncanny return, is a critical component of the ongoing creation of self.

If the mathematical encounters over time enable a fantasy of union and loss of self, but do not quite lead to fulfillment, then the potential is for one's life to become at least partially a quest for a completion of the fantasy. But if the classroom encounters are continually more likely to encourage a fantasy of harmonious resonance, a dream of students and teachers together merging through the mathematical quest, then the teacher will return again and again to the facilitation of classroom life, and find only perpetual iterations of disappointment in the inevitable returns that disrupt the dream. If the encounters lead to falling in love with mathematics, and the fantasy of sustained union beyond the encounters, then one will seek persistently to be in a relation with mathematics; When away, one will long to return to mathematics. Particular parts of mathematics are held in memory upon return from mathematical encounters, and will become treasured examples of what one loves about mathematics. Thoughts of mathematical objects and actions will be taken back out into the world as talismans against the vicissitudes of life. Anecdotal evidence from the world of schooling has suggested that virtually all pupils have already been forced by life to split off from their mathematical selves; schooling might be, but almost never becomes, a holding environment where the repair of such damage can take place (Winnicott, 1971, 1984).

What does mathematics demand of us? If we are to establish a relationship, we will always be both entering into a new encounter, and also projecting onto the relationship those characteristics of other relationships with others; by knowing and being known we have become, in a sense, the project that is ourselves. Yet we rarely stop to consider what mathematics might need (or want from), even as we use mathematics to work through those issues that are at the heart of our life project. As we pretend to merely "do" mathematics, we are at the same time negotiating a relationship *with* mathematics. So are our students, who may be persistently constructing such a relationship characterized by maintaining distance and separation, by projection of aspects of other relationships, and so on. Yet even as we try to understand our students as they develop interactions with mathematics, we may or may not be interpreting such interactions in ways that are more useful to our own ongoing construction of our own selves; and also as part of that (our relationship with mathematics), than to our understanding of our students (Appelbaum, 2008). Each time we attempt to create ourselves as teachers in some way that supports our dream of a teacher, we are also likely to be initiating either the violence of return or the encounters with mathematics that will eventually lead to the violence of return. So why do we and our students return, in the other direction, back to the mathematics?

Perhaps then, mathematics demands of us that we are in a Levinasian (1985) sense present to its face: we begin not with ourselves, but with mathematics as a radically ethical stance upon the world. Mathematics deserves from us that we not allow students and others to do it harm: to treat it in ways other than how mathematics itself would like to be treated; to respect mathematics for all that it has to give in response. Reifying mathematics and cutting it up into little bite-sized pieces, which can be memorized in a storage room of the mind to be exhibited later on display, would be one example of a particular form of violence upon mathematics; this might be akin to taking a human body and cutting it up into pieces for an exhibition on the human body. Just as the exhibit on the human body does little justice to the special experiences of that body when it is alive and living, so too would cutting up mathematics into small portable pieces be a disturbing and inappropriate approach. The paramount ethics begins with giving and serving, rather than the other way around; or for that matter, any other alternative to the relationship.

Therefore, we can see a possible (yet curious) reframing of "refusal" as "resistance." That is, students who might be interpreted as refusing to engage with mathematics, can at the same time be described as resisting such a relationship with mathematics. Yet even that is not adequate to the situation: students who might be described as "resisting" a relationship with mathematics, could perhaps be more interestingly re-interpreted as using resistance as a method for establishing a particular kind of relationship with mathematics. Long ago, Henry Giroux (1983) warned that we should be careful not to quickly brand any oppositional behavior as resistance; nor should we romanticize resistance as always having any sort of radical intentionality or the potential for social change. In the framework of critical theory, interpretations of resistance need to have their differences explored from among different kinds of resistance: those that embody a radical potential; those that work in opposition to hegemonic values yet, like the lads in Paul Willis' (1977) classic, *Learning to Labour*, ironically shut down social change; those that obscure critical consciousness; and, more subtly, those that educate about forms of injustice and inequity in ways that can further obscure or enable other forms of injustice and inequity. This sort of theorized resistance, leads to the promotion of education as emancipation: once students are made aware of oppression, and have been supplied with tools for overcoming oppression, resistance in a classroom might have to be interpreted as symptomatic of discontinuities in teaching and learning, which are not easily recognized with any theory of learning that imagines continuity between oppression and the experience of learning about oppression (Pitt, 2003). If learning disrupts the students' prior self-understandings, such disruptions, writes Alice Pitt (2003), can be experienced as producing new and often debilitating form of helplessness and isolation. This is why we need to frame resistance in the context at hand, in terms not only of the ways that acts of resistance are attached to the social, but also to the ways that individuals (teachers, learners, researchers, etc.) are attached to resistance. Falling in or out of love with mathematics pedagogically, suggests that we should ask something other than the usual questions about resistance (i.e., what constrains and enables political and social resistance, what potential acts of individual resistance hold for collective political action, and so on). A new layer of interpretation, or a parallel perspective, on the unfolding dynamics of mathematics education in local circumstances, can emerge from questions about how resistance and knowledge affect people; people who must negotiate social conditions that are not of their own making, but in which they find themselves.

Like a patient undergoing analysis, a student may exhibit any of the following: becoming unable to talk any longer; feeling s/he has nothing to say; needing to keep secrets from his/her teacher; withholding things from the teacher because s/he is ashamed of them; feeling that what s/he has to say isn't important; repeating him/herself constantly; refraining from discussing certain topics; wanting to do something other than talk; desiring advice rather than understanding; talking only about thoughts and not feelings; talking about feelings and not thoughts. (Appelbaum, 2008, p. 256)

At the same time, resistance shifts in the current framework, from a form of defense against scary, undesirable knowledge (of oneself, of the world, about others' intentions), toward a method of acting to produce specific kinds of knowledge (of oneself, of mathematics, of the teacher, of others...). The classroom, writes Pitt, "is implicated in the very social relations it hopes to observe, analyze, and disrupt." Resistance in this respect, "represents a mode of interpretive negotiation of the social as it is lived in classrooms." (p. 53) In the early sociology of education, resistance was observed where individuals confronted the otherness of ideology or its critique. In returning to (or from) mathematics, resistance takes place as a mode of relation when a person confronts the otherness of his or her unconsciousness knowledge. Furthermore, the stories I tell here about resistance and about classrooms, are forms of resistance used by my unconsciousness desire to know myself, and your reading of this is also a form of resistance toward the same ends.

Falling in love with mathematics, always yearning to be with mathematics, is a form of resistance method as well. Each of the behaviors listed in the previous paragraph, from "being unable to talk," to "talking about feelings and not thoughts," are also methods of being *with* mathematics; methods of yearning for the presence of mathematics, aching for mathematics, and so on.

> Falling in love is the ultimate act of revolution, of resistance to today's tedious, socially restrictive, culturally constrictive, humanly meaningless world.
>
> Love transforms the world. Where the lover formerly felt boredom, he now feels passion. Where she once was complacent, she now is excited and compelled to self-asserting action. The world which once seemed empty and tiresome becomes filled with meaning, filled with risks and rewards, with majesty and danger. Life for the lover is a gift, an adventure with the highest possible stakes; every moment is memorable, heart breaking in its fleeting beauty. When he falls in love, a man who once felt disoriented, alienated, and confused will know exactly what he wants. Suddenly his existence will make sense to him; suddenly it becomes valuable, even glorious and noble, to him. Burning passion is an antidote that will cure the worst cases of despair and resigned obedience. (CrimethInc, 2001)

So goes the dream of love that seems to be driving much of mathematics teaching. It certainly plays a part in my own planning, assessment practices, pedagogical decisions, and so on. The ex-workers collective claims this is very subversive, because it poses a threat to the established order of our lives. The boring rituals of workday productivity and socialized etiquette, they no longer mean anything to someone who has fallen in love; there are more important forces of guidance than mere inertia and deference to tradition. Yet, as C. S. Lewis (1960) admonished, "Of all arguments against love, none makes so strong an appeal to my nature as, 'Careful! This might lead you to suffering.'"[1]

> There is no safe investment. To love at all is to be vulnerable. Love anything, and your heart will certainly be wrung and possibly be broken. If you want to make sure of keeping it intact, you must give your heart to no one, not even to an animal. Wrap it carefully round with hobbies and little luxuries; avoid all entanglements; lock it up safe in the casket or coffin of your selfishness. (Lewis, 1960, p. 121)

The two perspectives on love are complementary, not in opposition.

PART II

Are mathematics as a body of practices, mathematical thinking as a collection of activities, and representations of mathematics objects of self rather than merely components of an academic discipline or a content to be learned? Might we imagine mathematical 'objects' parallel to a material world, in a discursive universe? If so, then mathematics might over time be both objects that people negotiate relations with in an ongoing experience of a becoming, emerging, constantly changing self. And mathematics and mathematical objects and mathematical practices would be 'things' through which people represent relationships with others and other, seemingly non-mathematical objects, the relations with which people negotiate as part of their ongoing life experience of becoming themselves (Tony Brown's *Mathematics Education and Language*, 1997). Here we have "objects" in a different sense from the historically paradigmatic, Platonic sense of ideal essences; such objects are static reifications, whereas the objects in this psychoanalytic framework are forever changing as ephemeral whisps of meaning, threads of relation, and metaphors that unconsciously guide people's actions. Jablonka and Gellert (2007) have written about mathematics as a world of "thinking abstractions" that take place in the imagination; yet also, as "implicit knowledge" that influences thinking and action independent of their initial purposes or the circumstances under which particular people had begun to reflect upon such abstractions (p. 7). More importantly, Stephen Brown (1984, 1993) has envisioned Platonic mathematics as redefined, and as an entity through which people can therapeutically reclaim their sense of self as a moral, acting being. For Brown, mathematics becomes transformed from a technique that links means and ends (a tool for "solving it"), into an activity through which one can understand oneself and see mathematics in new ways. In this view, the standard pole established by the skill drill versus meaningful conceptual knowledge, has been reframed as a persistent bypassing of activity that incorporates abstractions "out there in such a way that we can begin to gain power over it and feel that we possess it in some important sense" (Brown, 1993). "If we persist in by-passing this activity," wrote Brown:

> We desensitize ourselves to the point that we no longer taste the uniqueness among the phenomena, and though [students] may be able to gain answers to questions, they become very much insensitive to what it means for something to be a problem and have even less of an understanding of what it means to have solved something. (Brown, 1993, p. 271)

This insensitivity can be seen as a symptom of the "splitting crisis" or the "split off mind," which was referred to by D.W. Winnicott (1971), who earlier framed the issue of mathematical understanding in terms of object relations (Winnicott, 1986; Appelbaum & Kaplan, 1998).

Winnicott speculated in a 1968 talk for teachers, "Sum I Am," that there was a connection between an "integrated, unit self" (a personal sense of unity or oneness that emerges in infancy and early childhood through relationships between a child and its primary caregiver) and the development of a mathematical concept of "one" and a "unit." In his talk, he noted three types of object relations with mathematics, each of which might be characterized by the kind of unity, integration, disintegration, or un-integration experienced by the child as a response to the particular set of infantile and early childhood encounters. And we might infer from these types, school mathematics can come to represent or confirm these kinds of unity (or non-unity), as forms of object relations that are projected onto mathematical objects and practices. One way in which unit status would not be achieved, would be in a clear distinction between "me" and "not me." Such a split apart self, has no way of creating a meaning of "one," and we can infer from Winnicott's brief sketch that this type of self does not make much progress in mathematical tasks. Another type may also have failed to develop a personal sense of oneness, yet forges ahead to manipulate mathematical concepts despite being limited by trivial considerations of the unit concept; this type may engage in higher mathematics procedures, yet remain disconnected from understanding basic unit concepts. Such a condition is fairly common and typifies the person who has not achieved unit status, because as a child their environment required an application of intellect too early. This person may function brilliantly without reference to the human being, but has developed a false self (in terms of living with a split-off mind) "so that while higher mathematics gets a boost, the child fails to know what to do with one penny." (Winnicott, 1986, p. 59) The third type described by Winnicott, however, can relate to mathematics with an easy conception of oneness. This individual has a sense of personal unity, which has been derived from an experience with a good-enough behavior from a caregiver. This child's feeling of "I am" is available to be invested in a wider concept of wholeness, and helps build this person's personally relevant object relationships with mathematics.

If mathematics is thought of as a collection of tools through which people work through their ongoing forms of integration, disintegration, and unintegration, then, with Stephen Brown, we can indeed work to create opportunities for sensitivity and for tasting of the uniqueness of phenomena. For this to occur, the teacher should attempt to orchestrate the classroom as a holding environment in which object relations can be made and unmade, recreated and refashioned, and in which repair and reintegration can take place. It is in this sense that the experience of "return" would not be a "problem to be solved," but it would be an experience of value: so valuable the student would want to return again and again in the very same way with no change that might be labeled psychosis. To return (both to mathematic and away from mathematics) in new ways, and to use those moments of return as objects of self, becomes possibly the most critical dimension of classroom experience. How do we return? How do we feel as we return? What does it mean to return, and to turn again and again to returning itself?

Indeed, as Lacan proposed, a psychoanalytic session should be relatively short, since the work of psychoanalysis takes place mostly *between* sessions. He was fond of ending sessions abruptly, as a means of intensifying the questioning effect that in itself emphasizes a lack of closure. To borrow from that context, mathematical encounters might best be ended abruptly and in the middle of tangled up confusions, in order to make sure that the time in-between is most likely to be a time of important psycho-mathematical "work."

How would students react? In any of a myriad of ways. Each would in some sense reflect the students' relationships with mathematics, authorities (e.g., parents and their image of a good teacher), themselves as a learner, and so on. In much of my own teaching, students have used classroom encounters to seemingly affirm that they are good students; I inferred that "good" demonstrated a keen ability to follow directions and satisfy the goals of authorities. Their sense of self and their sense of knowledge and authority were entangled in mutually knotted ways. An abrupt interruption of an intense mathematical encounter, and the accompanying sense of disorientation, might have been met with bewilderment and disbelief, since good behaviour was expected to be rewarded with a perpetuation of the relationship with authority, that is, new instructions to follow if one is good. If the instructions were replaced with questions, these questions would be identified as an alternative to appropriate, good rewards. A learner might be modeled by a psychoanalytic client: he or she entered the experience seeking certainty and conclusion; this was met with a time period of getting lost and searching for interpretation. A time of concluding was artificially imposed on human experience, replacing anticipated certainty with entirely new questions to be pursued. In the end, there felt like there was no end to the experience. Analysis could last a lifetime. So can mathematics education.

Who is perpetrating this violence of return? That is, who does the student, with all of her or his experience in projections of relations with knowing and authority, perceive as the source of their anxiety and bewilderment? A student might blame herself or himself, the teacher, the subject matter, and so on. Or they might understand "return" as something upon which blame could not be the most obvious reaction; the latter experience was an end more clearly understood as a new beginning.

PART III

Questions about mathematics education have often been critiqued in terms of social and ideological contexts (Jablonka & Gellert, 2007). Interpretations of mathematics educational practices can often be discussed in terms of ideological outcomes and social reproduction (Jablonka, 2007). Yet, regardless of the structure of the curriculum or the pedagogy employed, the experience of return has been more or less orchestrated and negotiated without guidance in most schools, regardless of nation or level. Because students have had little experience with returning to the return itself, and therefore with understanding return as an all-at-once experience, a content and a method, there have been few students, if any, who ever encounter therapeutic mathematics. Few teachers have an intimate relationship with therapeutic mathematics; few mathematics education researchers are even aware of the potential for therapeutic mathematics. Indeed, therapeutic mathematics is likely to remain marginal and invisible in the professional field.

On the other hand, there has been a tradition in the United States and elsewhere of establishing a "thinking classroom" (Appelbaum, 2008), which carries the representation of mathematical thinking into mathematical literacy (Jablonka, 2003) and its codification as curriculum. In such evolving classrooms, teachers have noted that each small break from routine—whether a four-day weekend for Thanksgiving, a recess over the winter holidays, or even a day away from mathematics to attend a school assembly program—seems to require that students begin again their transformation from expectations of direct instruction, to interactions based on sharing ideas and offering conjectures. This form of repeated return to the clash between passivity, and the development of a language about one's thinking, has been a variation on the kinds of return discussed above, and might offer a glimpse into what can be achieved in bureaucratized forms of education.

Teacher education has the potential experience of return, as older students now re-experience their own earlier encounters; displaced in both time and location from the very classrooms in which they themselves practiced forms of return, resistance, splitting, and repair. In this context, "knowing mathematics" has become "knowing mathematics for teaching" (Ball, 1988). Overcoming the strong urge to "tell," can carry new meaning for those learning to teach within a psychoanalytic framework, which commonly has warned about the dangers of interpretation and imposition of narrative. Accepting cycles of return, have yet another feature of what Deborah Ball (1988) named "unlearning" to teach mathematics.

Another form of return can take place in classrooms that focus on "taking action" with what they learn (Appelbaum, 2009). Here students might disrupt ongoing learning experiences to ask themselves which audiences outside of their class do they interact with that relate to what they have been learning and doing. The revisiting of what has been experienced and the subsequent redesign of those experiences for others, can become a form of return "through the eyes of others"; it can be seen as a curriculum structure and a pedagogical method (learners creating an exhibition that brings the public into dialogue with them, designing social events based on recently learned material, etc.).

Finally, return as a node of mathematics education research and practice "returns us to the institutional and social" on a powerful, micro-level: it can become an evolving and constantly becoming "self" that becomes impossible to fathom without understanding such an individual as socially constructed, with multiple labels and forms of self-identification. This can include racial categories and interracial relationships; religious and ethnic affiliations and concomitant cross-group community experiences; sexual and age identities in varying contexts, and so on. None of this can be ignored as one imagines a mutually evolving macro and micro level of analysis.

> As curriculum conceptions often represent ideological hybrids, the consequences of mathematics curricula for different student groups in terms of their access to mathematical knowledge, their formation of mathematical identities and their positioning in the "knowledge society" are rarely directly visible. However, these consequences are not simply more or less accepted side effects of the practice of schooling. They reflect a differential distribution of legitimate and valued forms of knowledge and position, intended to reproduce or develop social structures. (Jablonka & Gellert, 2007, p. 6)

It is impossible to untangle the mathematical from the evolving (mathematical) self, society, tools, citizenship, cultural acculturation, and enculturation. Even as we begin to believe in an interpretation of "return," we will return to the problematic nature of reflection upon mathematics education theory, practice, and reflection itself: mathematics and our projections of mathematical objects as both tools of thinking and as characteristics of our relationships with ourselves, others, and mathematical events. These are sprinkled with ever-exponentially expanding repercussions of mathematics itself, as varieties of conflicting practices of social and individual construction and recreation. Eva Jablonka used a similar argument in a chapter on mathematical literacy (2003). She demonstrated that any attempt to promote a concept of mathematical literacy, was implicitly or explicitly a promotion of a particular social practice. In that chapter, she noted that critically evaluating aspects of one's surrounding culture was already affected by this culture as it was colonized by practices that involved mathematics; the ability to understand and to evaluate these practices, whether as a learner of mathematics, a teacher of mathematics, or a researcher observing mathematics education events, would of necessity then, form a recursive component of one's evolving mathematical literacy.

This evolving mathematical literacy has been in need of repair, as we simply work with and through mathematics education practices. Each return has the problem of return; each return could initiate return as the problem of return.

ACKNOWEDGEMENT

The author would like to thank Charoula Stathopoulou for her helpful questions on the original manuscript.

NOTE

1. As the Ramones say in their song, *Daytime Dilemma, Dangers of Love,* "The dangers, it's the dangers of love."

REFERENCES

Appelbaum, P. (2008). *Embracing mathematics: On becoming a teacher and changing with mathematics.* New York, NY: Routledge.

Appelbaum, P. (2009). Taking action: Mathematics curricular organization for effective teaching and learning. *For the Learning of Mathematics, 29*(2), 38–43.

Appelbaum, P. (2011). Carnival of the uncanny. In E. Malewski & N. Jaramillo (Eds.), *Epistemologies of ignorance and studies of limits in education* (pp. 221–239). Charlotte, NC: Information Age.

Appelbaum, P., & Kaplan, R. (1998). An other mathematics: Object relations and the clinical interview. *Journal of Curriculum Theorizing, 14*(2), 35–42.

Ball, D. L. (1988). Unlearning to teach mathematics. *For the Learning of Mathematics, 8*(1), 40–48.

Brown, S. I. (1984). The logic of problem generation: From morality and solving to de-posing and rebellion. *For the Learning of Mathematics, 4*(1), 9–20.

Brown, S. I. (1993). Mathematics and humanistic themes: Sum considerations. In S. Brown & M. Walter (Eds.), *Problem posing: Reflections and applications* (pp. 249–278). Hillsdale, NJ: Lawrence Erlbaum.

Brown, T. (1997). *Mathematics education and language: Interpreting hermeneutics and post-structuralism.* Dordrecht, the Netherlands: Kluwer.

CrimethInc. Ex-workers collective (2001). *Join the resistance: Falling in love.* Retrieved from http://crimethinc.com/texts/selected/joinresistance.php

Giroux, H. (1983). Theories of reproduction and resistance in the new sociology of education: A critical review. *Harvard Educational Review, 53*(3), 257–293.

Jablonka, E. (2003). Mathematical literacy. In A. J. Bishop, M. A. Clements, C. Keitel, J. Kilpatrick, & F. K. S. Leung (Eds.), *Second international handbook of mathematics education* (pp. 75–102). Dordrecht, the Netherlands: Kluwer.

Jablonka, E. (2007). The relevance of modelling and applications: Relevant to whom and for what purpose? In W. Blum, P. Galbraith, H.-W. Henn, & M. Niss (Eds.), *Modelling and applications in mathematics education: The 14th ICMI Study* (pp. 193–200). Berlin, Germany: Springer Verlag.

Jablonka, E., & Gellert, U. (2007). Mathematisation–demathematisation. In U. Gellert & E. Jablonka (Eds.), *Mathematisation and demathematisation: Social, philosophical, and educational ramifications* (pp. 1–18). Rotterdam, the Netherlands: Sense.

Levinas, E. (1985). *Ethics and infinity.* Pittsburgh, PA: Duquesne University Press.

Lewis, C. S. (1960). *The four loves.* New York, NY: Harcourt.

Pitt, A. (2003). *The play of the personal: Psychoanalytic narratives of feminist education.* New York, NY: Peter Lang.

Ramone, J., & Rey, D. (2002). "Daytime dilemma, dangers of love." in *Too tough to die,* (expanded) [compact disc], Los Angeles, CA: Rhino Records: 8122-78158-2.

Willis, P. (1977). *Learning to labour: How working class kids get working class jobs.* New York, NY: Columbia University Press.

Winnicott, D. W. (1971). *Playing and reality.* London, England: Routledge/Tavistock.

Winnicott, D. W. (1984). Aggression and its roots. In C. Winnicott, R. Shepherd, & M. Davis (Eds.), *Deprivation and delinquency* (pp. 84–91). London, England: Tavistock.

Winnicott, D. W. (1986). Sum, I am. In D. W. Winnicott (Ed.), *Home is where we start from: Essays by a psychoanalyst* (pp. 55–64). New York, NY: W.W. Norton.

Peter Appelbaum
Arcadia University, Philadelphia
USA

CHAPTER 2

BEYOND THE REPRESENTATION GIVEN

The Parabola and Historical Metamorphoses of Meanings

Christer Bergsten

INTRODUCTION

As well known, René Thom suggested that mathematics education should be founded on meaning rather than rigour. As if one could not have both. Having neither would not be an attractive option. Of course, it all depends (doesn't it always?) on what definition is given to these terms (meaning and rigour), and what mathematics education goals are being adhered to. One also needs to interpret the claim made by Thom through its context—a presentation at ICME2 in 1972 about "modern mathematics" during the time of the "new math" and its emerging problems:

> *The real problem which confronts mathematics teaching is not that of rigour, but the problem of the development of "meaning," of the "existence" of mathematical objects.* (Thom, 1973, p. 202, italics in original)

Refractions of Mathematics Education, pages 15–47
Copyright © 2015 by Information Age Publishing
15

The formalism of modern mathematics is not *natural* in education, Thom argues, but necessary only in numerical and algebraic computations; and the importance of rigour in mathematics is overestimated. Meaning is linked to consciousness and existence of the mathematical objects in the mental world. Contrasting algebraic language with Euclidean geometry, Thom finds that the move to

> eliminate elementary geometry to make room for calculus and linear algebra, has little to recommend it psychologically, because the algebraic objects (the symbols) are too poor semantically to permit themselves to be understood directly as is the case with a spatial figure. (pp. 207–208)

By its power to express structure,

> the spirit of geometry circulates almost everywhere in the immense body of mathematics, and it is a major pedagogical error to seek to eliminate it. (p. 208)

In contrast to a kind of secondary meaning *given* to algebraic symbols, geometrical objects appeal *directly* to intuition, one could argue (possibly in line with Thom), by their figurative appearance to the senses, providing a diagrammatic structure on which meanings can be built. By deleting elementary (or classical) geometry from the school curriculum, Thom argues, the "spirit of geometry" that also enters mathematical reasoning when the focus is not geometry itself, would jeopardize the presence of meaning generally in mathematics.

Now, the problem of meaning did not disappear with the new math. Formal mathematics, in the sense of numerical and algebraic computations, fill up much of students' work in elementary mathematics classrooms today, for many coupled with a double meaninglessness: the symbols used (as Thom noted), but also the rules by which they are operated on. This chapter, though, will ask the question: what happens to this "spirit of geometry" (in terms of the meaning it provides) with its *algebraization*, which turns it into a study of coordinates and graphs of functions. Consider the parabola, for example: it is not only problematic, according to Bartolini Bussi (2005), to define a conic section algebraically, "it is not possible to build the meaning of conics through only a one-sided approach, as, for instance, through the most widespread algebraic definition." (p. 39). In line with this quote, it would also be impossible to remain within the Euclidean geometric system of reasoning tools when aiming for *the* meaning of conics. If there is such a thing. Always lurking behind any discussion of meaning in mathematics, and its education, is Russell's (1901) comment that mathematics, "may be defined as the subject in which we never know what we are talking about nor whether what we are saying is true." Such statements stem from a specific philosophy of mathematics, an unavoidable link also in education as Thom (1973) reminds us about: "In fact, whether one wishes it or not, all mathematical pedagogy, even if scarcely coherent, rests on a philosophy of mathematics." (p. 204)

As a contribution to the discussion of meaning in mathematics education, this chapter will draw on the example of the parabola to investigate, through a historical lens, the relation between representations and meanings of a mathematical object. After a short introductory theoretical account on this relation and some mathematical preliminaries, a commented search for the life of the parabola through the history of mathematics will be presented. This will be followed up by a discussion of the elusive notion of meaning. In the final part of the chapter some issues related to meaning in the teaching of the parabola in school mathematics will be raised, drawing on the notion of praxeology (or mathematical organization).

The parabola and the other conics have been studied extensively in the history of mathematics, as well as in mathematics education. Is there really more to say? Indeed, this chapter will also draw on an earlier presentation (with the same title as this chapter[1]) in a discussion group at the PME conference in Bergen, Norway in 2004, as well as some parallels in Bartolini Bussi (2005). However, the exercise in this chapter aims to elaborate in some detail on different representations of a specific mathematical object (the parabola), in a search for some refractions of the relation between meaning and rigour in school mathematics.

REPRESENTATIONS AND MEANING

Thom's claim about the necessity of focusing on meaning rather than rigour in mathematics education (with reference to a mental existence of mathematical objects), invites some reflections on meaning. Indeed, conceptions of meaning rest on "presuppositions concerning the relationship between the cognizing subject and the object of knowledge" (Radford, 2006, p. 40). The ontological issue about whether this *object of knowledge* refers to something that has an existence independent of the knower, provides an essence of objectivity to the knowledge necessary to earn it a status of knowledge that can be shared with others and not exclusively a purely subjective construction. This has been discussed also within mathematics education. Radford (2006), for example, made the distinction between an *objective referential approach*, "which locates mathematical objects outside semiotic activity" and "played an important role in the shaping of the pedagogical scene at the turn of the 20th century" (p. 41), a *discursive approach* in which mathematical objects are constituted as discursive objects within the mathematical discourse (Dörfler, 2004; Sfard, 2008), and *a semiotic-cultural approach* where "ideas and mathematical objects…are conceptual forms of historically, socially, and culturally embodied reflective, mediated activity" (Radford, 2006, p. 42) with an emphasis on the role of *semiosis.* The latter approach described *objectivity* as contextual and cultural, in which "the relationship between the observer and that which is observed is a culturally mediated one" (Radford, 2006, p. 60), where "the whole arsenal of signs and objects that we use to make our intentions apparent" constitute the *semiotic objectification of knowledge* (Radford, 2003). This is put in contrast to Peirce's view where ultimately the true objectivity resides at the horizon, beyond an endless process of semiotic chaining, steered by his *pragmatic maxim;* and where "meaning is the relation that links one sign to the next sign (i.e., its interpretant) in a semiotic chain" without taking into account social praxis and human interaction (Radford, 2006, pp. 44–45). For Radford then, meaning is both a subjective and a cultural construct, linked to the individual's personal history and experience but also "prior to the subjective experience, the intended object of the individual's intention (l'object visé) has been endowed with cultural values and theoretical content that are reflected and refracted in the semiotic means to attend to it" (Radford, 2006, p. 53).

Representations are also in the center for Damerow (2007) in his account for the connection between historical developments and individuals' cognition, when outlining a theory of the historical development of arithmetical thought. "[B]asic structures of logico-mathematical thought," he writes,

> are developed by the individual growing up in confrontation with culture-specific challenges and constraints under which the system of action have to be internalised. The challenges are embodied in the material means of goal-oriented or symbolic actions that are shared representations of logico-mathematical structures of mental models. (p. 22)

He makes a distinction between first- and second- (or higher) order external representations of mental models. Here the *first-order representations* are:

> Material representations of real objects by symbols or by models composed of symbols and rules of transformation, with which essentially the same actions can be performed as with the real objects themselves (p. 25)[2]

Among the examples provided, constructions with compass and ruler are first-order representations of geometrical figures (in the Euclidean plane), useful for locating objects, their relations and movements. Regarding the function of first-order representations, actions on the representations could be performed more easily and freely than actions on the objects themselves; to foresee the results of the real actions. These symbolic actions also "initiate the construction of the same mental models as actions with the real objects they represent" (p. 26). While the same symbols are used in different contexts, their meanings become differentiated. To represent mental models, however, second- (or higher-) order representations are needed, consisting of

> symbols or models composed of symbols and rules of transformation, which correspond to the operations of the abstract mental model that controls the actions performed with the real objects. (p. 26)

To be used properly, a second-order representation needs to be related to the real objects and actions that are to be represented by its symbolic system by "assimilating them to the mental model that gives the external representation its meaning" (p. 26). For example, some theorems in Euclidean geometry (e.g., the Pythagorean theorem) are second-order representations, as they do not relate to concrete geometric figures, but to discursively defined mathematical objects within a deductive framework. Through second-order representations meta-cognitive mental models "may be constructed as higher-order representations by reflective abstraction" (p. 27). As the external (material) representations are historically transmitted, they can be used for the reconstruction of the mental models (reflected meanings) they represent.

In school mathematics, students constantly encounter and are expected to construct meanings through work with exercises on historically transmitted representations of mathematical objects. Here the use of higher-order representations is ubiquitous but often without established links to lower-order representations and their "cultural values and theoretical content," thus leaving potential gaps in the students' meaning-making processes.

Why Look at the Parabola?

Consider the following claim, having in mind its relevance for meanings that students may experience in their mathematics studies: In today's school mathematics, concepts and methods are often treated in isolation, with only surface level theoretical embedding within a mathematical subdomain. This very general statement, which is endowed with a considerable degree of specification, is not supported here by any reference or argument; nevertheless, it is taken as the starting point for this chapter.

The treatment of concepts and methods, at least in the Swedish secondary school context that will be referred to here, is more or less dominated by algebraic tools. However, as often pointed out, many students do not achieve well in algebra, in neither of the aspects mentioned above. As a consequence, an image of mathematics as disconnected and difficult to understand may emerge in students. Applications used in teaching to enhance motivation then easily get the character of decorations rather than parts of an integrated knowledge structure. One example that illustrates this didactic phenomenon is the parabola. Problems and techniques related to this classical curve are ubiquitous in school mathematics:

- The concept of square root, often used as a basis for the extension of the number concept from rational to real numbers
- The solution of quadratic equations, often used as a basis for the extension of the number concept from real to complex numbers
- A prototype for polynomials and the connection between zeros and factoring of polynomials
- Second degree polynomials often serve as a prototype for the study of derivatives and optimisation problems
- A prototypical example for mathematical modelling, e.g., of reflections (parabolic antennas) and projectile motion

Evidently, a student's conceptualisation of a parabola is framed by how it is described, defined, treated, or applied in the educational context. The term *concept image* has been coined to capture the (non-static) mental outcome of this process (Tall & Vinner, 1981), indicating the "existence of the mathematical objects in the mental world." A key role is here taken by the representations used, and how the mathematical object in focus is being objectified by their mediation (Radford, 2003). In today's school mathematics, different representations of a mathematical object (such as the parabola) all live in a mixed world of mathematical ideas and tools, and are not always treated by a systematic approach, as seen from the students' perspective. Here the asymmetry (in specialized knowledge) between the teacher and the student comes into play, as the teacher may have links between mathematical ideas (implicitly) in mind, while presenting a specific mathematical object, ideas not available to the students. The parabola, with its long cultural history, is such an object.

Preliminaries

In order to find an initial meaning for parabola, I will provide a short reminder of some preliminaries that are based on Figure 2.1, drawing on two elementary mathematical tools: the Pythagorean Theorem and mean proportion. I will focus on how some basic properties of the two-dimensional curve named parabola can be derived from its original three-dimensional construction from cutting a right-angled cone by a plane parallel to its surface.

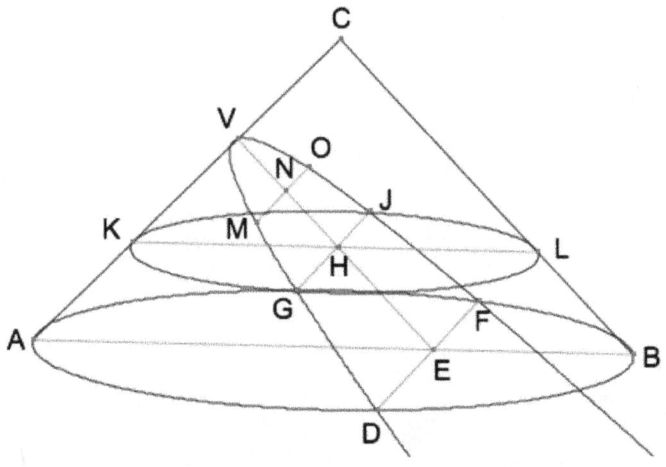

Figure 2.1   A parabolic cut from a right-angled cone.

In Figure 2.1 AB is the diameter of the circle produced by a plane cutting a circular right-angled cone (with vertex C and right angle ACB) perpendicular to its axis. The section DVF (the *parabola*) has been produced by a plane cutting the cone parallel to BC with DF perpendicular to AB. If the distance CV is of length a, then HL and EB are both of length $a\sqrt{2}$. The point H should here be imagined as moving along VE. In the circle with diameter KL, GH is the mean proportion of KH and HL. Thus, with GH having length x and VH length y, KH has length $y\sqrt{2}$, and it follows that $x^2 = y\sqrt{2} \cdot a\sqrt{2}$, i.e., $x^2 = 2ay$. In this way, the quadratic function is intrinsically linked to the parabola.[3]

Now consider the specific value of x for which x and y are equal (GH = VH), i.e., $x^2 = 2ax$. The non-zero solution $x = 2a$ is in Figure 2.1 drawn as the segment DE, and MN has half this length ($x = a$). Again, by mean proportion in the circle at 'level' N (not drawn in Figure 2.1), one finds that $VN = \frac{a}{2}$, which is a fourth of DE, and also of VE. The point N is of special interest considering the distance d from the arbitrary point G, on the parabola, to N (see also Figure 2.2):

$$d^2 = GH^2 + HN^2 = x^2 + \left(\frac{x^2}{2a} - \frac{a}{2}\right)^2 = \frac{1}{4a^2}(x^2 + a^2)^2$$

It follows that $GN = VH + \frac{a}{2}$. This is the well-known *equidistance property* of the parabola, illustrated in Figure 2.2 where the plane containing the parabolic section in Figure 2.1 is laid out.

In Figure 2.2, VN′ is constructed equal to VN and thus, as shown above, GN = HN′ = GG′. The point D was chosen so that ED = EV and M chosen so

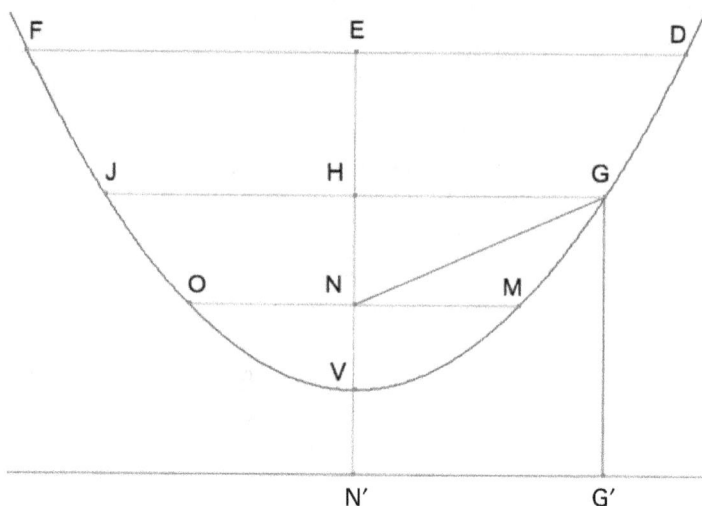

Figure 2.2 The equidistance property of a parabola.

that OM = ED. The chord OM, passing the *focus* N, is often referred to as *latus rectum* ('straight line') and its length (equal to 2*a* in Figure 2.1) as the *parameter* of the parabola. As seen above, the length of latus rectum is four times the distance from the focus to the vertex. The line through N′ and G′ is known as the *directrix*.

HISTORICAL ORIGINS

Tracking the history of a mathematical object reveals a double influence on its development—from inside and outside the pure mathematical realm. In this part of the chapter, I will provide a short account of the historical progression of the mathematical object parabola, within classical geometry, analytic geometry, and dynamic geometry software. These steps will parallel those described by Bartolini Bussi (2005, pp. 40–42) for the historical development of conics in mathematics and society. From a semiotic point of view, a chain of metamorphoses of meanings will be described. My interest is on how this becomes critical as the mathematical object, by the social process of didactic transposition, is turned into an object for teaching and learning.

Archimedes

In the works of Euclid, Archimedes, and Apollonius the parabola is a geometric object, defined rhetorically, and analysed by the tools of constructive

and deductive (Euclidean) geometry. Historically, the conic sections first appeared, it is believed, in the attempts by Menaechmus (approximtely 350
B.C.) to solve the problem of duplicating the cube (Eves, 1983, p. 80). Later,
in his famous treatise on the quadrature of the parabola, Archimedes speaks
of the parabola as "a section of a right angled cone" (Archimedes, 1952,
p. 527)[4] and refers to "the elementary propositions in conics which are of
service in the proof," three of which are referenced to in works by Euclid
and Aristaeus. One of these "elementary propositions" listed is of special interest in a discussion of meanings related to the parabola (see Figure 2.3):[5]

> If from a point on a parabola a straight line be drawn which is either itself the
> axis or parallel to the axis, as PV, and if from two other points Q, Q' on the
> parabola straight lines be drawn parallel to the tangent at P meeting PV in V,
> V' respectively, then PV: PV' = QV²: Q'V'².

This proposition is thus by Archimedes taken as "proved in the elements
of conics." However, it appears later in Apollonius' *Conics* along with a proof
(see the section on Apollonius below).

Here, the reader might wonder about the purpose of this chapter; obviously placed in the educational domain, but now elaborating on a (rather
standard) historical *Rückblick*, perhaps asking as did Fried (2013, p. 2),
"is there really a distinct educational relationship to our mathematical
past?" A rationale for a mathematical elaboration of this historical part of
the text for mathematics educators, apart from the intended discussion
of meaning, is that this piece of mathematics may have some unexpected
relevance when re-contextualised from the educational point of view. For
example, the proposition quoted might be an eye-opener, which questions
the privilege given to the orthogonal Cartesian coordinate system in school

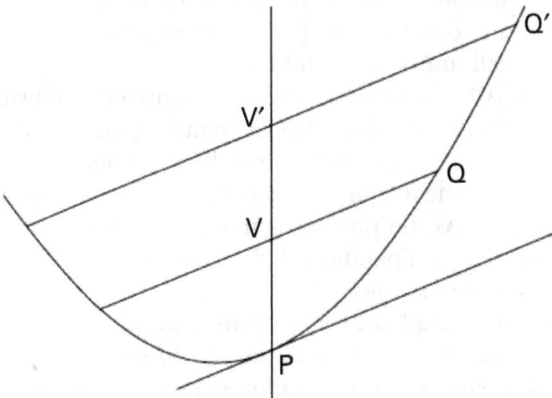

Figure 2.3 An illustration to a proposition referred to by Archimedes (1952, p. 528)

mathematics and its formatting over the imagined range of mathematical relationships. In the paragraph that will follow, however, the relevance is presented as inherent in the proposition itself; this is done by way of bridging different mathematical domains, thus providing a potential counteract against the fragmentation of school mathematical knowledge highlighted in the introduction. As a clarification of this claim, by allowing a semiotic jump when interpreting the proportional equality $PV : PV' = QV^2 : Q'V'^2$ as an algebraic equation of quotients, the introduction of the variables $y = PV$, $y' = PV'$, $x = QV$, $x' = QV'$ in the expression brings it into the form

$$\frac{y}{y'} = \frac{x^2}{x'^2}.$$

This implies that

$$\frac{y}{x^2} = \frac{y'}{x'^2} = k$$

and thus $y = k \cdot x^2$. It should be noted here that the lines PV and QV are not necessarily perpendicular.

Application of Areas—A Reconstruction

The well known proposition referred to by Archimedes relates the parabola, defined as a conic section, in modern mathematical terminology and within another mathematical domain, to the quadratic function. The initial constructions involved in relating the parabola to a square, in line with the proposition above, may have been built on the Pythagorean method of applications of rectilinear areas.[6] As an illustration of this method within the present context, the construction (or reconstruction) presented in Figure 4.2 is offered, which draws on Euclid I.43.[7]

In Figure 2.4, AB is a given segment (parameter). Drawing on Euclid I.43, an application of the square AEGD (standing on the variable segment AE) to a rectangle standing on AB will produce the point P (as the intersection point of EG and the diagonal AC) and the rectangle ABFH (with equal area as the square). As the point E is moving on AB or its extension the point P will move along a parabola. Using algebraic notation, with $AE = x$, $EP = y$, and $AB = p$, we have $py = x^2$.

The previous paragraph seems to be an attempt to "revoice" the mathematics of the past (Boero, Pedemonte, & Robotti, 1997) in a way that makes the reader recognize the common parabolic icon, which is vertically placed on orthogonal coordinate axes. However, as the method of application of areas works equally well in the non-orthogonal case, Figure 2.5a

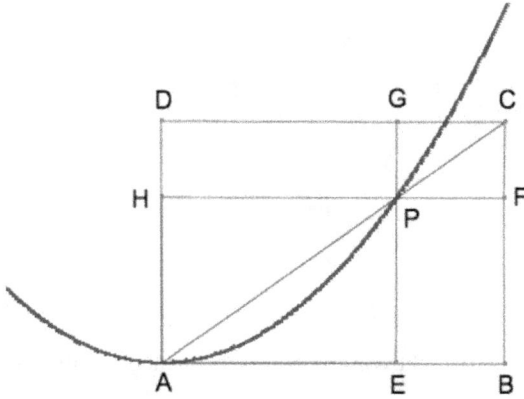

Figure 2.4   A construction of a parabola by the application of areas of rectangles (Bergsten, 1998, p. 170).

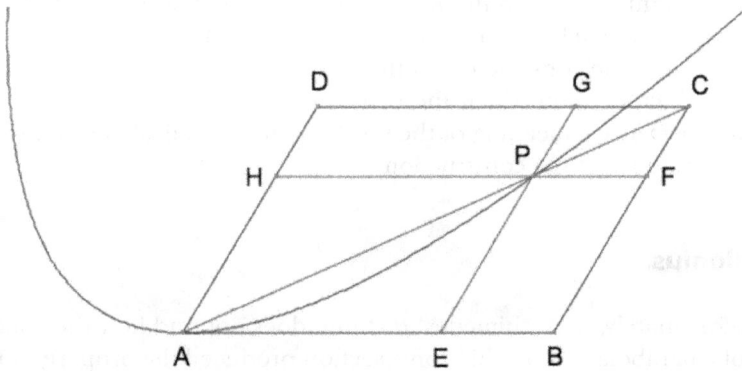

Figure 2.5a   A construction of a parabola by the application of areas of parallelograms.[9]

better illustrates this interpretation (meaning making) of the proposition mentioned by Archimedes. When the rhombus on AE is applied on the parallelogram on AB (here the length of AB is a fixed parameter), the parabolic curve appears as the locus of the point P, as E is moved along AB or its extension. This construction of the parabola thus draws on the property of a parabola stated in the proposition mentioned previously by Archimedes.[8] Figures 2.4 and 2.5a were constructed using geometry dynamic software employing the drag mode and trace function to display, in modern mathematical terms, the functional relationship between the independent variable (AE) and the dependent variable (EP, equal to AH). The parameter AB shapes the parabola as more or less open.

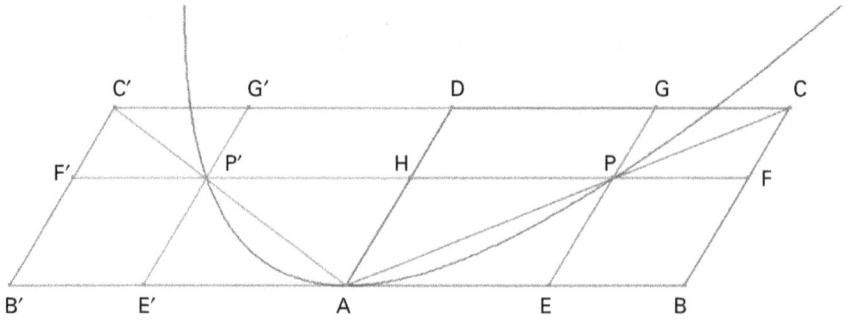

Figure 2.5b Symmetry in the construction of the parabola by the application of areas.

For the left part of the parabola in Figure 2.5a, Figure 2.5b illustrates the symmetric construction (where AB′ equals AB, and AE′ equals AE). By the drag mode in a dynamic geometry software this leaning parabola can be straightened up to the rectangular case (such as in Figure 2.4) by dragging the line defining the upward direction of the parallelograms (see Figure 2.8 and the construction of the focus).

The principle upon which the construction draws can make the individual experience a meaning of the resulting mathematical object through the performance of the construction.

Apollonius

Unfortunately, in Archimedes' text one does not find how the conception of a parabola as a specific conic section produced the property (proposition) quoted above and illustrated in Figure 2.1, as the work referenced is lost. This, however, one can find in Apollonius' *Conics*; he defines a diameter of a section as a line passing through the midpoints of parallel chords, and the axis of the section as the diameter that is perpendicular to parallel chords. His definition of parabola, which is contained in a proposition with some similarities to the one given above, "is long because more than a statement of a claim, it is the description of a diagram" (Fried, 2007, p. 216). This certainly is true also for his definition of a cone (Apollonius, 1952, p. 604). Apollonius' definition of a parabola is as follows (Figure 2.6):

> If a cone is cut by a plane through its axis, and also cut by another plane cutting the base of the cone in a straight line perpendicular to the base of the axial triangle, and if further the diameter of the section is parallel to one side of the axial triangle, then if any straight line which is drawn from the section of the cone to its diameter parallel to the common section of the cutting

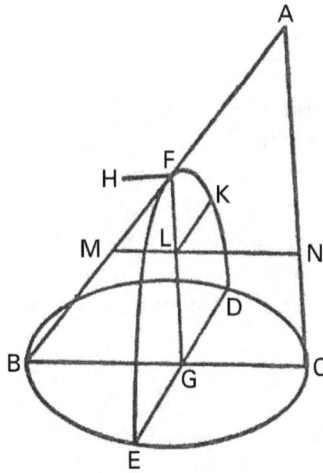

Figure 2.6   Apollonius' definition of a parabola (Apollonius, 1952, p. 615).

plane and of the cone's base, will equal in square the rectangle contained by the straight line cut off by it on the diameter beginning from the section's vertex and by another straight line which has the ratio to the straight line between the angle of the cone and the vertex of the section that the square on the base of the axial triangle has to the rectangle contained by the remaining two sides of the triangle. And let such a section be called a parabola. (Apollonius, 1952, p. 615)

With reference to Figure 2.6 (where the diameter FG is assumed to be parallel to AC), Apollonius defines the parameter FH (perpendicular to FL) by relating it to the shape of the cone through the requirement $FH : FA = BC^2 : BA \cdot AC$.[10] He then proves, by the use of elementary Euclidean propositions on triangle similarity, the proposition $LK^2 = LF \cdot FH$, corresponding to the application of areas shown in Figure 2.4 (where the parameter AB has replaced the parameter FH in Figure 2.6, and been chosen independently as there is no cone involved; different lengths of AB would correspond to different vertex angles of the cone). From this proposition (Proposition 11 in Conics I), the proposition referred to by Archimedes and illustrated in Figure 2.3 immediately follows (as Proposition 20 in Conics I).

To Apollonius' proof, Calinger (1995) makes the editor's note that, "It was Apollonius's most distinctive achievement to have based his treatment of the conic sections on the Pythagorean theory of the *application of areas*" (p. 165; italics in original). This was however not a defining property as it follows a "then." The focal property was not discussed for the parabola but might be inferred as limiting cases from such discussions about the hyperbola and the ellipse (Apollonius, 1952, p. 788). The terminology used, based on drawings where the size of segments are supposed to vary, have been seen by some as predecessors of both the coordinate system of analytic geometry and the concepts of variable and parameter as used in the study of functions (Charbonneau, 1997; Figure 2.4); this is common with a gaze on history from modern mathematics. As a historian refraining from employing such gaze, Fried (2007) noted that "the parabola, for Apollonius, is essentially a geometrical object and it is tied to the cone" (p. 216). However, as seen above (in quote from Calinger, 1995, p. 165), by observing the application of areas involving a square, the parabola could leave the cone and continue its life independently in a two-dimensional world; even more so after the discovery of the equidistance property.[11] That way, with Apollonius the meaning of a parabola as a purely geometric object can be anchored in the cone, by its construction as a section, or in the application of areas by the identification of a square and a rectangle within the plane of the parabolic curve. A new basis for a potential *change of meaning* appeared again later with Pappus: the meaning of a parabola could now be anchored on the equidistance property, still within the same representational register of figurative configurations that can be drawn or visualized as a basis for reflection and reasoning. Borrowing the terms *treatment* and *conversion* from Duval (2006), a treatment within the same semiotic register might incur adjustment or change of meaning, while a conversion to another register would imply a metamorphosis[12] of meaning. The latter could also be seen as paradigmatic shift: that will be the topic for the next section.

To arrive at the first historical metamorphosis, a time travel of some 1850 years is now made from one line to the next (a jump also made by Bartolini Bussi, 2005, in her historical account of "the meaning of conics"), creating a gap in the coherence of the textual exposition as it leaves out the long-term developmental process of the semiotics of the mathematical signs involved. This, however, does not affect the arguments made. It is also claimed that after Pappus "no significant progress in the study of conic sections had been made until the work of Kepler" (Koudela, 2005, p. 198).

A PARADIGMATIC METAMORPHOSIS

After the advent of analytic geometry, most notably in the (independent) work of Descartes and Fermat, where the more modern algebraic notation in Descartes' book *La géomètrie* from 1637 was the most influential, a line (curve) could be described by the relationship between coordinates (referring to distances along given directions) and studied by algebraic treatment. By representing a point on a parabola by an algebraic equation of the coordinates at the point, it was, by a historical metamorphosis, transformed from a geometrical object into an algebraic object: from being described by a rhetoric sequence (referring to a configuration), often along with a diagram (drawing), it now showed itself as an algebraic expression such as $y = x^2$, still though, along with natural language. By defining the geometrical properties of a curve, it was possible by the use of coordinates to find an algebraic representation of this curve,[13] provided that the algebraic manipulations could be coped with. As an example, from the defining property for a parabola as the loci for points with the same distance to a fixed point (*focus*) as to a fixed line (*directrix*),[14] an algebraic calculation can provide the equation $x^2 = 4ay$ if the focus point is in $(0, a)$ and the directrix is $y = -a$, using a current standard ON system. As the algebraic equation $y = x^2$ (or, with the use of set theory formalism, $\{(x, y) \in R^2 : y = x^2, x \in R\}$) does not in itself constitute an icon of the curve it generates by its interpretation into coordinates for points in a standard coordinate system, its status as a representation of a parabola is abstract, as expressed by Fried (2007, pp. 216–217):

> To understand our own mathematical approach to objects such as the parabola, then, it is essential to see how Descartes relocates mathematical thought from the system of ideas of Apollonius' world to a new system of ideas in which objects like the parabola are redefined and reinterpreted in terms of abstract relations (i.e., relations lifted up and away from special subject matter).

By this "relocation" of mathematical thought into a new "system of ideas," a historical metamorphosis of meaning took place. For example, when the parabola in a school curriculum is introduced algebraically as a quadratic function, the meaning linked to the curve being more or less narrow or wide is "relocated" to the size of the coefficient of the quadratic term, sustained with a computational argument (for the same value of x, $3x^2 < 4x^2$) prior to its interpretation in a diagram. In the conic section definition, the increasing wideness can be directly perceived as the cut with the plane that is done further away from the vertex of the cone (a bigger parameter), where its circular base is wider.

The fact that Fermat reverted this process by finding the geometric counterparts starting with given algebraic equations (Eves, 1983, p. 265), also shows how the power of representations through semiotic invention and chaining may produce new and unexpected knowledge. Curves never before thought of, such as the parabolas of Fermat (given by the equations $y^n = ax^m$), or in modern time fractals, could be described by algebraic means and interpreted iconically when represented in the plane (or space) by coordinates.[15] Descartes performed in the second book of his *Géomètrie* a classification of all possible curves produced by a quadratic equation in two variables, and opened the way for higher order equations. His analytic method is clearly expressed in the following lines:

I could give here several other ways of tracing and conceiving a series of curved lines, each curve more complex than any preceding one, but I think the best way to group together all such curves and then classify them in order, is by recognizing the fact that all points of those curves which we call "geometric", that is, those which admit of precise and exact measurement, must bear a definite relation to all points on a straight line, and that this relation must be expressed by means of a single equation. If this equation contains no term of higher degree than the rectangle of two unknown quantities, or the square of one, the curve belongs to the first and simplest class, which contains only the circle, the parabola, the hyperbola, and the ellipse; but when the equation contains one or more terms of the third or fourth degree in one or both of the unknown quantities (for it requires two unknown quantities to express the relation between two points) the curve belongs to the second class; and if the equation contains a term of the fifth or sixth degree in either or both of the unknown quantities the curve belongs to the third class, and so on indefinitely. (Descartes, 1954, p. 48)[16]

Curves corresponding to the equation $y^n = px$, $n > 2$, for example, are considered parabolas of higher order (Eves, 1983, p. 264). New meanings may also be created: when the properties of geometric objects are studied by work in the algebraic domain, the latter may serve as a means to better understand, or gain new knowledge, of the former (Bolea, Bosch, & Gascon, 1999). The symbolic language of algebra can thus be seen as a didactic tool to learn about the other domain, in this case geometry, which is fundamentally different (Bergsten, 2003).

As an example of Descartes' algebraic work as he performed the classification of the second-degree equation, the following short passage gives a flavour of both his skill and enthusiasm for his new method:[17]

Again, for the sake of brevity, put

$$-\frac{2mn}{z} + \frac{bcfgl}{ez^3 - cgz^2}$$

equal to o, and

$$\frac{n^2}{z^2} - \frac{bcfg}{ez^3 - cgz^2} \text{ equal to } \frac{p}{m};$$

for these quantities being given, we can represent them in any way we please. Then we have

$$y = m - \frac{n}{z}x + \sqrt{m^2 + ox + \frac{p}{m}x^2}$$

(Descartes, 1954, p. 63)[18]

These quotes serve as warrants for the use of the term historical metamorphosis, even though the previous paragraphs are unhistorical in any other sense than by their presentation as an account of appearances of certain texts at different points of time in the history of mathematics. The focus is on the metamorphosis of meanings, those residing beyond the representations given, and that are in Peircean terms within the interpreter. In the case of *les anciens*, a term Descartes frequently used as a reference for Greek mathematics during antiquity, a meaning of the sign parabola (παραβολή) could be inferred from a verbal description of a geometric construction. While the close affinity between the different conics has its origin in a gradual change within the construction (the angle of a cutting plane in relation to a cone, or the size of the lack or excess of an applied area) for *les anciens* and Descartes it is a matter of the gradual change of size of coefficients in a quadratic algebraic equation in two variables.

The historical metamorphosis of geometry into analytic geometry was paradigmatic. As soon as the sometimes heavy geometric analysis had been complemented with smooth algebraic calculus, the development of mathematics exploded. Not only the ancient conics, now called quadratic or second-degree curves, but also corresponding 3-D second-degree surfaces[19] could be systematically studied (for example by Euler, who introduced coordinate transformations to find the canonical form representation; see Kline, 1972, pp. 545–547), also in the general setting of quadric forms, which by matrix notation and eigenvalue theory were given a unified and systematic treatment during the 19th century (Kline, 1972, pp. 799–812). The formal notation of analytic geometry has been easy to expand into any dimension, as shown by the first printed publication that focused on higher-dimensional point geometry by Cayley in 1843 and other early independent work by Grassmann and Schäfli (Eves, 1983, p. 415).

The next historical metamorphosis of the meaning of the parabola that will be discussed in this chapter is linked to the development of digital computational tools in the 20th century.

A COMPUTATIONAL METAMORPHOSIS

A cone can be physically realized in wood, and cut by a plane realized as a metal saw, leaving the cut surface open for observation. The shape of the contour, which has been produced by the cut, is being born; and with visual perception and reflection, it takes on its own life as a cultural object (Radford, 2006), a 2-D curve: the parabola. Whether such a birth of the parabola happened according to the standard Menaechmus story or not, or in the writings of Apollonius, by such type of intersection of two 3-D objects (a cone and a plane in space), a 2-D object was constructed and named παραβολή (Greek for application, here of areas) along with an investigating of its geometric properties (some of which have been highlighted previously). However, while being developed by means of technology (even if those 3-D objects were only mentally conceived or drawn, that is by a mental technology), the new product (the parabola) became and has remained an intellectual object for over two millennia; it is in essence static as represented by drawings on paper or as algebraic equations.[20] Controlled and systematic animation required a computational technology that did not (yet) exist.

With the advent of computer technology in the 20th century, performing high speed numerical calculations paired with advanced screen technology, a user could not only plot a graph from a given algebraic formula, but also could study geometrical properties of the graph (as for example a parabola) as (seemingly connected) points on a screen by doing simple hand movements using the "drag mode" of a dynamic geometry software, to realize animation while keeping construction generated invariances intact. The geometric (iconic) object is again in focus, and this time directly available without the intermediate representation of an algebraic formula, subject to manipulation like the algebraic equations implicitly used to describe it and visualize it on the screen. By a historical metamorphosis, the parabola can now appear as a *dynamic object* on a computer screen. Mathematics has given itself, with its advanced technology, one more didactical tool to possibly better understand itself.

However, before appearing and being animated on the screen, the parabola studied has to be defined as a specific parabola, either by a formula or a geometric construction performable by software tools: Its meaning has to be *anchored*. The semiotic means of objectification (Radford, 2003) thus draws heavily on the cultural history of mathematics by its transmitted material representations, as well as technological artifacts. That the dynamic features of these new representational tools (such as dynamic geometry software) can impact meaning can be seen in the construction of the parabola presented in Figure 2.4. Using the drag mode, a user can *experience* how point P is pressed down toward the horizontal axis when point E approaches A, and pushed up when E is moved beyond B. Applying the equidistance construction of the parabola (Figure 2.7), the perpendicular bisector used for the construction appears as a tangent to the parabola, and the reflection property can be immediately conceived by the drag mode, other invariants not mentioned. The construction shown in Figure 2.4, as well as the standard one in Figure 2.7, are both easy to perform (provided the software has been instrumented) and can generate to learners self-steered and controlled experiences on which meanings of the parabola that are lurking beyond the representation used, can be constructed.

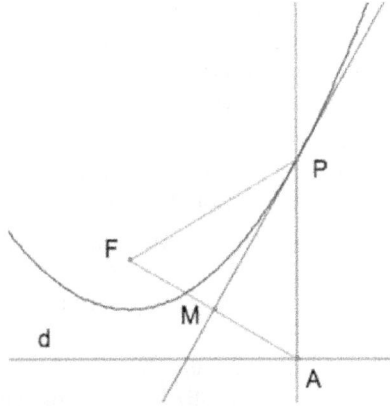

Figure 2.7 Equidistance construction of a point (P) on a parabola.[21]

DISCUSSION

The section produced by cutting a cone along a plane is, in Damerow's (2007) terms, a first-order representation of a real object. With some variation of how the cut is performed, an individual may develop a mental model of a curve called a parabola. The establishment of properties, such as those outlined in the preliminaries above, involve second- (or higher-) order representations, while employing deduced theorems from circle geometry. When some of these properties are used for constructing (defining) the parabola, another mental model of the curve is evoked, from which new higher-order representations can develop. With the historical transmission of these representations, a reconstruction and development of related meanings within the semiotic register of Euclidean geometry has been possible. However, with the intermediate representation of distances as algebraic symbols (the method of Descartes), the superimposition of a coordinate system on the parabolic curve created a conceptual blend that constituted a paradigmatic metamorphosis of meaning. A higher-order representation of a geometric curve as an algebraic equation, can refer to the original object only indirectly, via an intermediate higher-order representation, requiring a chain of connected mental models to have meaning.

The paradigmatic historical metamorphosis of the parabola pictured previously, may be conceived as an algebraization of a mathematical organization (or work; see the section on school mathematics that follows): "a mathematical work is algebraized if it can be considered as an algebraic model of another mathematical work, the system to be modelled" (Bolea et al., 1999, p. 142). Such kinds of modelling should respond to all the

techniques and technological questioning in the organisation being modelled as a whole (Bolea et al., 1999). This seems to be part of what Descartes
aimed at with his analytic method (Descartes, 1954, p. 48), but it also took
the source organization further and his method proved self-generating for
the development of mathematics.

By the "parabolic" cases in 2 and 3 dimensions, i.e., $y = x^2$ and $z = x^2 + y^2$,
the obvious (in some sense) "parabolic" extension to n dimensions would
be $y = x_1^2 + x_2^2 + x_3^2 + \ldots + x_{n-1}^2$, where $\bar{x} = (x_1, x_2, x_3, \ldots, x_{n-1}) \in R^{n-1}$. From a concrete action on a physical object—cutting a cone—the iconic "section" of a
parabola[22] by a historical metamorphosis returned in the shape of a symbolic representation in the semiotic register of algebra. By the development of
this register, hitherto unknown mathematical objects were hidden beyond
transformation, variation, and generalization of this representational form.
With the new eyes, new things were found, and old things were looked at
in new ways. As a consequence of this semiotic relocation and because of
the human quest for meaning, when images from sensual experience are
not available (as in the case of multidimensional "geometry" in R^n, $n > 3$),
the phenomenon of metaphorical mapping (Lakoff & Núñez, 2000, p. 43)
comes into play; this is witnessed by the choice of names and diagrams most
often produced for communication. Examples include the expression "unit
ball" when referring to the set $\{x \in R^n : x_1^2 + x_2^2 + x_3^2 + \ldots + x_n^2 < 1\}$, or the use
of the word "distance" between functions as elements in a Hilbert space
(Bergsten, 1993).[23] A similar claim was made in Bourbaki (1974) about the
increased abstractness of mathematics, "to the extent that geometry has
been transformed into a universal language for contemporary mathematics" (Bartolini Bussi, 2005, p. 40). Is this an indication of the impact of the
"spirit of geometry" on meaning-making in mathematics?

By the computational metamorphosis the iconic representation of a geometrical object turned dynamic. Although the individual can experience it
as a direct control of the real object, the object on the computer screen is
recomputed by the software by each touch of the drag mode; thus higher-
order representations are produced. That the object is only virtually real as
a material higher-order representation does not prevent it from support-
ing mental models that carry meaning. The dynamic feature of the virtual
object allows the discovery of invariances by the use of drag mode.[24] The
virtuality of the object is irrelevant to the proof of the invariance of Euclidean geometry's discursive tools.

Is it then meaningful to ask what a parabola *is*, beyond the representation given? Is there a truth hidden at the horizon, in line with the Peircean
dream? What is it that holds the concept of a parabola together, or what is,
in Husserl's (1989) terms, the *ideal object* parabola? What seems to remain
invariant across the representations given, is the iconic configuration (Bartolini Bussi, 2005, p. 40). But the "spirit of geometry" cannot reside in this

configuration per se, but in the elaboration of the geometric properties embedded in this configuration through its mode of construction: embedded relative to the discursive part of a given mathematical organization (such as Euclidean geometry, that René Thom may have had in mind). Where the real object beyond the representation resides, if anywhere, seems to have no relevance for the experience of meaning. What has taken place has been a chain of re-descriptions by way of chains of higher-order representations and reconstructions of mental models.

The semiotic perspective shows how meaning is transferred through the representational forms to what the interpreter, within that person's given cultural situation and background, constructs from them. And the historical development of mathematics shows, as illustrated here by the example of the parabola, how cultural meanings can change its face when new semiotic registers are developed and used on objects that used to be studied by other registers. Through a *semiotic chain*, the intersection of a cone with a plane (an icon) has been linked to a quadric form (the notation of which is a symbol), represented by a symmetric matrix (symbol), and to electronic dots on a computer screen (together constituting an icon); this constitutes a move from first-order to second- and higher-order representations. However, new developments do not always replace old meanings with new ones, but rather they grow as new layers of signification outside the old. A process of a similar kind, at the individual level, has been described by Presmeg (2006) by a Peircean nested triadic model of semiotic chaining. The model, resonating with Radford (2006), may be applied also to a historical-cultural development of meaning, as one of the parabola sketched previously. The development of new representations by this process is not expected to end at some unified conception (in this case) of what a parabola is. In analyzing the historical development of another mathematical object, the fundamental theorem of calculus, Jablonka and Klisinska (2011) concluded that one would expect not a convergence, but even more diversity of meanings; and that it is in education rather than within the scientific discipline that efforts of unification are found.

An additional concept of Peirce may overarch, and in the realm of reasoning and problem solving integrate the iconically based argumentation of Euclidean geometry, as well as the symbolically based in analytic geometry, that is, diagrammatic reasoning:

> All deductive reasoning, even simple syllogism, involves an element of observation; namely, deduction consists in constructing an icon or diagram the relations of whose parts shall present a complete analogy with those parts of the object of reasoning, of experimenting upon this image in the imagination, and of observing the result so as to discover unnoticed and hidden relations among the parts. (Peirce, quoted in Dörfler, 2004, p. 7)

Also the manipulation of algebraic formulas is a work with icons, the patterns of formulæ: "These are patterns, which we have the right to imitate in our procedure, and are the icons par excellence of algebra." (Peirce, quoted in Dörfler, 2004, p. 7; cf. the concept of *mathematical form* in Bergsten, 1999). This kind of diagrammatic reasoning is of such generality that it may function, for the individual as a linking force between different mathematical domains, since it is not primarily focused on the referential function (or meaning) of the forms (representations).

THE PARABOLA IN SCHOOL MATHEMATICS

As well known, René Thom suggested that mathematics education should be founded on meaning rather than rigour. As if one could not have both. The German tradition of *Stoffdidaktik* within mathematics education, seems to have aimed at precisely this with its emphasis on grounding teaching on mathematical background theories, ways of making fundamental ideas available to students at different age levels, and *Grundvorstellungen* in students (Tietze, 1994; Bergsten, 2014). Principles used in the process of selection and analysis of the mathematical content to be taught include elementarizing, exactifying, simplifying, and visualizing (Tietze, 1994), as well as complementarity of theoretical and pragmatic aspects of knowledge (Otte & Steinbring, 1977). The present discussion has tried to highlight the complementarity of meaning and rigour through an elaboration of historically transmitted forms of external representations as semiotic means of objectification, and has drawn on the analyses of Radford (2006) and Damerow (2007). Reference has also been made to the *Anthropological Theory of the Didactic* (ATD; Chevallard, 1998), which will form the basis for the discussion of school mathematics in this section.

In this theoretical approach (ATD), two inseparable aspects of mathematical activity within an institution are identified: one practical (know-how) and one theoretical (know-why). The former consists of types of tasks or problems that are studied and the techniques used to solve them; while the latter is formed by the corresponding discursive environment, that is, issues of technology (the discourse and ingredients related to the techniques) and theory (deeper justification), together constituting mathematical organizations or *praxeologies* (Chrevallard, 1998; Bosch & Gascon, 2006). The ATD grew out of earlier elaborations on didactic transposition theory (Bosch & Gascon, 2006).

For example, the task of finding the midpoint of a given circle may be solved by the technique of constructing midpoint perpendiculars to two chords in the circle; thus using technological ingredients such as chords to circles, justified by theorems in Euclidean geometry. Mathematical organizations can be considered at different levels, where a *punctual* mathematical organization of a specific type of problem and technique that is used to solve it, can be embedded in a *local* mathematical organization with the technology available for using those techniques. Some local mathematical organizations may build on the same theoretical discourse to form a *regional* mathematical organization (Bosch & Gascon, 2006). By the process of didactical transposition, different levels of co-determination (i.e., the level of society, school, pedagogy, discipline, area, sector, theme, and subject) set didactic constraints on classroom activity. The classroom activity of a teacher is more or less restricted to the last two levels, with the mathematical scope of the three preceding levels (Bosch & Gascon, 2006).

Given this theoretical framing, students' first acquaintance and work with the parabola is constrained by the local mathematical organizations they have previously met, with their integration or fragmentation, and which mathematical organisation it is embedded in. Levels of justification also vary with the didactic transposition induced by the curriculum, and with the kinds of problems studied; the meanings made available to students can thus vary accordingly.

Examples of Didactic Transpositions of the Parabola

For the upper-secondary science student in Sweden in the 1960s, a parabola was the geometric locus of points equidistant to a given point (*focus*) and a straight line (*directrix*), a property which was immediately elaborated by the algebraic register of analytic geometry to the formula $y^2 = 4ax$, upon which a systematic treatment of a number of geometric and algebraic properties was based (Sjöstedt & Thörnqvist, 1963, pp. 62–76).[25] The study of the parabola was embedded in a local mathematical organization of analytic geometry, including techniques of Euclidean and coordinate geometry, as well as elementary algebraic equation solving. Techniques from the study of functions, such as the derivate, were not used.

Later, in the 1980s and 1990s, a metamorphosis of meaning had been introduced: for the same student the parabola was now introduced as an algebraically defined curve of second degree, $y = x^2$, the shape of which was plotted, based on a table of values (e.g., Björk, Borg, Brolin & Ljungström, 1990, pp. 264–274). There was little or no discussion about any geometric properties of the curve (such as equidistance or reflection), other than the symmetry (based on the algebraic argument that $(-x)^2 = x^2$) and the vertex point. In applied textbook tasks, specific algebraic equations for the curves involved were mostly provided. Tangents and optimization problems were then treated by means of the derivative. The study of the parabola was embedded in a local mathematical organization of functions. Techniques from Euclidean geometry, as well as analytic geometry, were used only casually, apart from the basic idea of coordinate representation as a means for plotting the graph of a function. The textbook by Bergendahl, Håstad and Råde (1975) marks a transition state where the parabola is introduced as a second-degree polynomial curve (embedded in a mathematical organization of functions) in the main text, but accompanied by an appendix (optional) chapter about the conic sections framed within analytical geometry.

For university students, shortly, the acquaintance with quadratic surfaces (such as the paraboloid) and the corresponding plane curves, is accomplished by the study of quadratic forms, embedded in the domain of linear algebra. However, the application of integrals on rotational volumes for the same mathematical objects, is often done in isolation from linear algebra within the local mathematical organization of calculus (based on a discursive technology involving real numbers, functions and limits of functions).

Objectification of Knowledge

In many educational situations, the student is presented objects of learning previously not known (or encountered), as might be the case with the parabola in a mathematics class. For the student to become aware of this object, the teacher might use "semiotic means of objectification—e.g., objects, artifacts, linguistic devices and signs that are intentionally used by the individuals in social processes of meaning production" (Radford, 2002, p. 14). The drawing of a parabola may seem the optimal such means, but is by itself void of meaning beside its appearance. It is necessary to go beyond the representation given to find meaning. The same remark also holds for an algebraic representation, such as $y = x^2$ or $y^2 = 4ax$, of the parabola. In the case of the icon, a defining property such as equidistance (focus-directrix), does not only give the shape of the icon (the curve), but also provides a basis for further explorations and analyses of the object as a support for attaching meaning to it. Using another defining non-algebraic property, such as the application of areas as outlined in Figure 2.4, will generate the same shape but offer another basis for further explorations, and meanings:

> The figural representations of conics...are invariant in time and hence not subject to historical changes. What are changed are the way of generating conics, the way of looking at them, and the way of studying them. (Bartolini Bussi, 2005, p. 40)

The objectification process for the learner diverges between the options, even more so by providing an algebraic formula as the main means of objectification; this leads into a different mathematical organization by the different techniques available. In a learning setting, such options to relate to other representations or other semiotic registers are restricted by the overall didactic situation; this is as a consequence of the didactical transposition that has taken place. By those constraints, students' work may remain at the punctual level, resulting in concept images often too vague for successful problem solving in situations spanning over more than the restricted punctual or local mathematical organization that has been covered in class.

An Example of Pedagogy

With reference to work of Vygotsky and Bachtin, among others, Boero, Pedemonte, and Robotti (1997) have designed an "educational situation" they call the "voices and echoes game," which seems to be akin to the theoretical outlines by Radford (2006) and Damerow (2007) on the relationship between the individual and the socio-historical knowledge development, with a specific focus on the role of representations in the meaning making processes. By encountering historical approaches in texts that represent "historical leaps" in the development of mathematics, students can produce *echoes* to these *voices* from the past through work on specific tasks. These echoes can be individual or collective, depending on the organization of classroom activities. The aim of the activity is to find ways to organize the mediation and appropriation of theoretical knowledge.

Historical metamorphoses of meanings (as have been discussed in this chapter), may be useful as bases for the design of such specific tasks, as they can provide semiotic means of objectificating mathematical objects. What may be called bridging tasks,[26] will link well to the present discussion, and will provide the techniques and technology of the different mathematical organizations that are available to students. Such tasks would create a bridge between (historically based) different representations of a mathematical object (a conversion between different semiotic registers or a treatment within the same semiotic register); it could be a bridge that would have the potential to carry meanings across representations and mathematical organizations. An obvious example related to the parabola, would be to show that the curve defined algebraically by the equation $y = x^2$ (where x and y are Cartesian coordinates for points on the curve), has the equidistance property, and vice versa. An equally obvious follow up from this task would be the demonstration of the reflection property of a parabola in the different mathematical organizations related to these two definitions. In the case of the algebraic definition, the technology of the mathematical organization for the study of functions, may include the derivative. Here, meanings available in the geometric case may provide a background gaze to the technicalities of the algebraic work needed to solve the task within the study of functions algebraically defined; the landscape of meaning would remain barren without feeding it with the spirit of geometry.

In a bridging task involving a comparison of the two geometric constructions of the parabola shown in Figure 2.4 and Figure 2.7, respectively, the representations can be both within the same semiotic register. One question to ask in such comparison would be whether the parabola produced by the application of areas in Figure 2.4 has the equidistance property on which the construction of the curve in Figure 2.7 is based (to justify the use of the name parabola also for this curve). Leaving this "pleasure of discovery" to the reader, a similar but more comprehensive bridging task will instead be discussed here in more detail:

> **Example:** *How can one find the axis and the focus point of a parabola constructed by the application of areas as in Figure 2.5a?*

By this example a bridge can be established between the parabola as defined by the equidistance property, where the focus point would be given, and as defined by the application of areas in the non-rectangular case in Figure 2.5a (that is within the same mathematical register and organization of Euclidean geometry). The example raises some critical issues on meaning and its dependence on a reference knowledge base.

Figure 2.5a does not itself offer much advice on how to proceed. However, Figure 2.5b shows that the line through A and D passes the midpoints of parallel chords to the parabola, and thus is a diameter to the parabola, parallel to the axis (drawing on Apollonius). This means that a chord PP′ perpendicular to this diameter can be drawn, which would also be perpendicular to the axis. Consequently, the perpendicular bisector to PP′ must produce the axis, meeting the parabola at its vertex V (see Figure 2.8 where MV is the axis). To proceed with the problem it is thus necessary to bring in additional theoretical knowledge, such as definitions and properties of diameters and the axis of a parabola, and knowledge about where on the axis the focus point is located.

As shown in the preliminaries above, the distance from the vertex to the focus point is one fourth of the parameter (equal to the length of the *latus rectum*). Select, then, the point B′ on the line through A and D so that AB′ is a fourth of AB. Using the drag mode of the software to straighten up the parallelogram ABCD to a rectangle, will show that the line through B′ parallel to AB meets the axis at the point F, the focus point of the parabola. Select F′ on the extended axis so that VF′ is equal to VF. The line perpendicular to the axis at F′ is the directrix to the parabola (with PF equal to PQ; the tangent at P to the parabola is the perpendicular bisector to F and Q). It can also be observed (using the drag mode) that the line through A

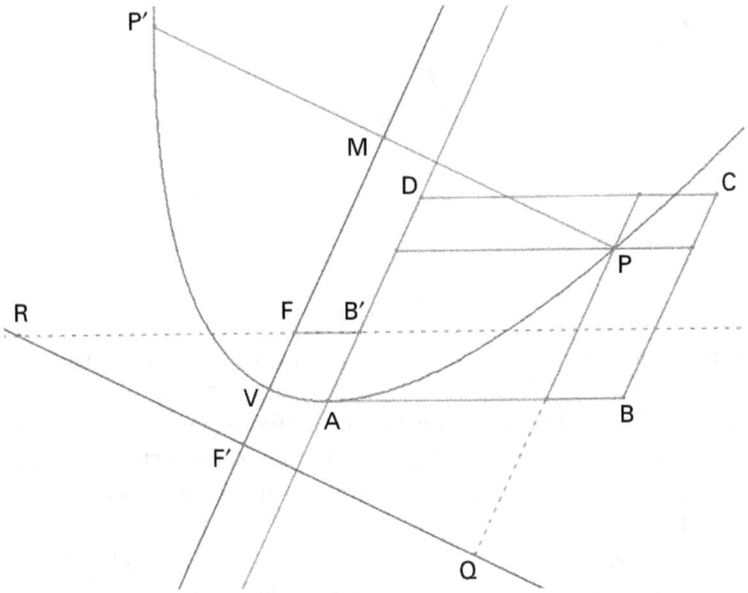

Figure 2.8 Construction of the axis and focus of a parabola as defined in Figure 2.5a.

and V meets the directrix, and the line through B′ and F, at their point of intersection R. Here, this property does not seem to add more meaning to the construction, though geometry invariance is always compelling. Using the drag mode (e.g., varying the angle of the parallelogram ABCD) other invariances can be discovered. For example, the circle with center at A and passing through B′, also passes through F and touches RQ. This means that both F and the directrix can be constructed simply by drawing this circle. Alternatively, the circle with center at B′ having a radius of twice the length of AB′ (and thus a diameter equal to the parameter AB), meets the parabola at two points defining a chord parallel to AB and passing through F.

The constructions above for finding F thus draw heavily on moving between theoretical and figural (or diagrammatic, in the sense of Peirce) reasoning, thus constituting an experience of meaning production (Laborde, 2005). What seems to work figuratively (confirmed by the use of drag mode), however, is not always theoretically evident; this implies the need for a theoretical search for meaning. In this case, the claim that F so constructed is, indeed, the focus point, is such an issue. This could lead to the search for meaning beyond other representations. One option to find F, draws on knowledge about the *latus rectum*: Construct the point S on the line through P and P′ so that the length of MS is twice the length of MV (see Figure 2.9). The line through V and S then meets the parabola at the endpoint T of latus rectum (by similarity), and the perpendicular to the

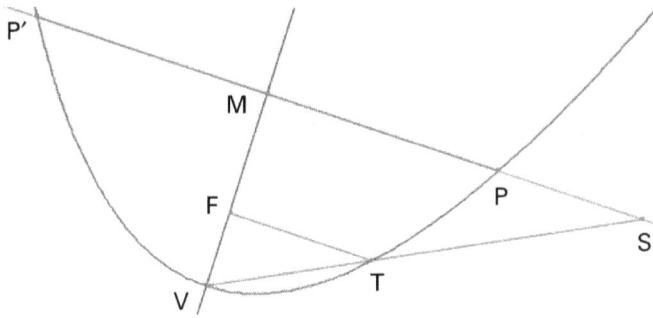

Figure 2.9   Construction of the focus point of a parabola with a given axis.

axis through T will thus meet the axis at the focus point F (producing the right half FT of latus rectum in Figure 2.9). This construction is potentially rich in meaning, but requires that the software can produce intersection points with the parabola.

Bridging tasks may also be designed to draw more explicitly on the history of mathematics, as in the teaching experiment described in Bartolini Bussi (2005) where historical mathematical machines are used for the construction of the parabola along with the pre-Apollonian definition of conics as sections of a right-angled cone; or as in examples of the "voices and echoes game" described in Boero et al. (1997).

Concluding Remarks

An individual involved in mathematical work today has access to potentially all historically transmitted material representations of a mathematical object. This holds, of course, for teachers and students at all levels of education, as well as educational policy makers. There is, therefore, always a choice when a curriculum is to be designed regarding which mathematical organization is seen as most appropriate in relation to the specific educational goals and socio-cultural conditions. It is a question of curriculum development to design school mathematics education to support students' development of conceptual models and problem solving techniques, so that their space of reference is not constrained to the punctual level. By linking the different representations of the objects of learning (e.g., the parabola) to more general mathematical organizations into which they can be embedded, the representational flexibility needed for solving more non-routine problems can be developed through the refractions of the complementarity between meaning and rigour that this evokes.

NOTES

1. The title alludes to Bruner (1973) who argued that "when one goes beyond the information given, one does so by virtue of being able to place the present given in a more generic coding system and that one essentially reads off from the coding system additional information either on the basis of learned contingent probabilities or learned principles of relating material" (p. 224).
2. This description parallels the notion of a *genetic mathematical form* (see Bergsten, 1999), referring to a symbolic expression that has a form (structure) isomorphic to a material realisation of the signified.
3. In terms of areas, the formula states that the square on GH equals the rectangle with sides VG and twice CV. As CV is a constant, a relation is thus found between VH and GH for an arbitrary point G on the parabola.
4. By this definition a parabola is the section produced by a plane cutting a right angled cone at a right angle to an element of the cone (cf. the preliminaries above).
5. The power notation on QV is used by the editor of Archimedes' text to refer to the square with each of its sides equal to QV. Figure 2.3 does not appear in Archimedes' text but a similar diagram with only one chord related to the following proposition: "If from a point on a parabola a straight line be drawn which is either itself the axis or parallel to the axis, as PV, and if QQ′ be a chord parallel to the tangent to the parabola at P and meeting PV in V, then QV = QV′." (Archimedes, 1952, p. 528)
6. See Sardelis and Valahas (2012). The method of application of areas is described by Proclus (ibid., p. 1). The names that Apollonius introduced for the conic sections (parabola, ellipse, and hyperbola) have their origins in such applications (Eves, 1983, p. 128).
7. An online version of the *Elements* is available at http://farside.ph.utexas.edu/euclid/Elements.pdf
8. The method presented here for the construction of (points on) a parabola differs from the "conceptual reconstruction" in Sardelis and Valahas (2012), which builds on the (standard) construction of a mean proportion in Euclid II.14. In terms of the mathematical language of functions, Sardelis and Valahas construct the *square root function* while Figure 2.4 (and Figure 2.5a) constructs the (inverse) *quadratic function* (though both by the application of areas). The reconstruction in Sardelis and Valahas (2012) is more aligned to the formulation in Apollonius (in connection to Figure 2.6), even though they have argued that their "conceptual reconstruction supports the view that the conics were most probably discovered as plane curves by fusing the method of the application of areas with the concept of locus, long before Apollonius studied them as conic sections" (https://scirate.com/arxiv/1210.6842). Their method also covers an application of areas for the other two conics (drawing on constructions in Euclid). It should be noted that the method used in Figure 2.5a, is based on the proposition presented in Archimedes' text and supports the same view, and can also easily be extended to the other two conics with the use of Euclid II.14 for one step of the construction. However, the margin is too small to present these constructions here.

9. This figure may be called a *parabelogram*. In the formula $y = k \cdot x^2$ for the parabola in Figure 2.5a, where AE = x and AH = y, one has $k \approx 0.2$. The size of the constant k does not depend on the leaning angle of the parallelogram but only on the length of the parameter AB (an increase of AB produces a decrease of k).

10. In Figure 2.6, FG is parallel to AC, MN is parallel to BC, and DE is perpendicular to BC.

11. That the parabola (as defined by Apollonius) has the equidistance property (to a *focus* point and a *directrix* line) was shown by Pappus (Kline, 1972).

12. In Merriam-Webster's online dictionary (http://www.merriam-webster.com/dictionary), metamorphosis refers to "a major change in the appearance or character of someone or something."

13. In the embodied cognition literature this phenomenon is termed a *conceptual blend* (Lakoff & Nuñez, 2000, p. 48).

14. In the *Conics* of Apollonius we find the corresponding properties of the hyperbola and the ellipse in propositions III.51 and III.52, respectively.

15. As well known, Descartes did not draw two perpendicular coordinate axes and used only positive coordinates. His *method* was made more available to the contemporary public by Van Schooten's 1649 Latin translation and commentary. A "modern" coordinate system with perpendicular positive and negative x- and y-axes and an origin O one finds for example in Newton.

16. It can be observed here how Descartes still sticks to the tradition when interpreting second order terms like xy or xx geometrically ("rectangle," "square"), but calls higher order terms more freely as "third or fourth degree," and writes them with exponential notation.

17. This enthusiasm also echoes in the very last words of the 1637 treatise: "I hope that posterity will judge me kindly, not only as to the things which I have explained, but also to those which I have intentionally omitted so as to leave to others the pleasure of discovery." (Descartes, 1954, p. 240)

18. Descartes used the notation xx for x^2 and a special sign for equality.

19. According to Kline (1972, p. 321), Fermat had already in 1643 anticipated coordinate descriptions of these 3-D surfaces.

20. The parabola also survived as an object of study in schools thanks to, I would hypothesize, its practical usefulness stemming from some of its geometric properties.

21. To the segment between the fixed point F (focus) and a variable point A on the fixed line d (the directrix), draw the perpendicular at the midpoint M, intersecting at point P the perpendicular to d at A. The point P is on a parabola by the equidistance definition. Note that the distance from F to d determines the shape of the parabola (though when scaled all parabolas are similar).

22. The name parabola, though, alludes to the idea of application of areas and not to cone cutting.

23. As a remark here, Apollonius' definition of a conic surface displays an image schematic character (the path schema; see Lakoff & Núñez, pp. 37–39, 141) in being thought-of generalised bodily based images.

24. One example of such 'discovery' is given in the discussion of Figure 2.8.

25. The approach of Apollonius was also included in this textbook, for optional deepening studies.
26. The label "bridging task" is currently used in different educational contexts, as for example literature studies where the aim is to link a text to its historical/social/cultural context.

REFERENCES

Archimedes (1952). The quadrature of the parabola. In *Great books of the western world, Vol. 11* (pp. 527–537). Chicago, IL: Encyclopedia Britannica.

Apollonius of Perga (1952). Conics. In *Great books of the western world, Vol. 11* (pp. 593–804). Chicago, IL: Encyclopedia Britannica.

Bartolini Bussi, M. (2005). The meaning of conics: Historical and didactical dimensions. In J. Kilpatrick, C. Hoyles, O. Skovsmose, & P. Valero (Eds.), *Meaning in mathematics education* (pp. 39–60). New York, NY: Springer.

Bergendahl, G., Håstad, M., & Råde, L. (1975). *Matematik för gymnasieskolan 2, NVT* [Mathematics for upper secondary schools 2, NVT]. Stockholm, Sweden: Biblioteksförlaget.

Bergsten, C. (1993, September 17–20). On analysis computer labs. In B. Jaworski (Ed.), *Proceedings of the international conference on technology in mathematics teaching (TMT93)* (pp. 133–140). University of Birmingham, England.

Bergsten, C. (1998). Mötespunkter [Crossroads]. In I. Olsson, E. Proffe, L. Sterner, & C. Edholm (Eds.), *Matematik som kultur. Dokumentation av 10:e Matematikbiennalen* [Mathematics as culture. Documentation of the 10th mathematics biennial] (pp. 169–174). Sundsvall, Sweden: Mitthögskolan.

Bergsten, C. (1999). From sense to symbol sense. In I. Schwank (Ed.), *European research in mathematics education I.II* (pp. 126–137). Osnabrück, Germany: Forschungsinstitut für Mathematikdidaktik,

Bergsten, C. (2003, February 27). A classification of algebraic tasks. Paper presented at *New trends in mathematics education research: international perspectives* seminar, Bologna, Italy.

Bergsten, C. (2004, December). Beyond the representation given: The parabola and historical metamorphoses of meanings. *SMDF Medlemsblad, Nr 10,* (pp. 37–49). Linköping, Sweden: SMDF. Retrieved from http://www.mai.liu.se/SMDF/arkiv/MB10

Bergsten, C. (2014). Mathematical approaches. In S. Lerman (Ed.), *Encyclopedia of mathematics education.* New York, NY: SpringerReference.

Björk, L-E., Borg, K., Brolin, H., & Ljungström, L-F. (1990). *Matematik. Lärobok NT1* [Mathmatics. Textbook NT1]. Stockholm, Sweden: Natur och Kultur.

Boero, P., Pedemonte, B., & Robotti, E. (1997). Approaching theoretical knowledge through voices and echoes: A Vygotskian perspective. In E. Pekhonen (Ed.), *Proceedings of the 21st international conference on the psychology of mathematics education, 2* (pp. 81–88). Lahti, Finland.

Bolea, P., Bosch. M., & Gascon, J. (1999). The role of algebraization in the study of a mathematical organization. In I. Schwank (Ed.), *European research in mathematics educationi I.II* (pp. 138–148). Osnabrück, Germany: Forschungsinstitut für Mathematikdidaktik.

Bosch. M., & Gascon, J. (2006). Twenty-five years of the didactic transposition. *Bulletin of the International Commission on Mathematical Instruction, 58,* 51–65.

Bruner, J. (1973). *Beyond the information given.* London, England: Allen & Unwin.

Calinger, R. (Ed.) (1995). *Classics of mathematics.* Upper Saddle River, NJ: Prentice Hall.

Charbonneau, L. (1996). From Euclid to Descartes: Algebra and its relation to geometry. In N. Bednarz et al. (Eds.), *Approaches to algebra* (pp. 15–37). Dordrecht, the Netherlands: Kluwer.

Chevallard, Y. (1998). Analyse des pratiques enseignantes et didactique des mathématiques: l'approche anthropologique [Analysis of teaching practices and the didactics of mathematics: An anthropological approach]. In *Actes de l'université d'été, La Rochelle, 4–11 juillet 1998* (pp. 91–120). Clermont-Ferrand, France: IREM.

Damerow, P. (2007). The material culture of calculation. A theoretical framework for a historical epistemology of the concept of number. In U. Gellert & E. Jablonka (Eds.), *Mathematization and de-mathematization: social, philosophical and educational ramifications* (pp. 19–56). Rotterdam, the Netherlands: Sense.

Descartes, R. (1954). *The geometry of René Descartes.* New York, NY: Dover.

Dörfler, W. (2004). Mathematical reasoning and observing transformations of diagrams. In C. Bergsten & B. Grevholm (Eds.), *Mathematics and language* (pp. 7–19). Linköping, Sweden: SMDF.

Duval, R. (2006). A cognitive analysis of problems of comprehension in learning of mathematics. *Educational Studies in Mathematics, 61*(1–2), 103–131.

Eves, H. (1983). *An introduction to the history of mathematics* (5th ed.). Philadelphia, PA: Saunders College.

Fried, M. N. (2007). Didactics and history of mathematics: Knowledge and self-knowledge. *Educational Studies in Mathematics, 66,* 203–223.

Fried, M. N. (2013, March). The varieties of relationships to mathematics of the past. [Paper presented]. HPM-Americas, United States Military Academy at West Point, NY. Available at http://www.hpm-americas.org/meetings/2013-east-coast/

Husserl, E. (1989). The origin of geometry. Translated by J. P. Leavey, Jr. and reprinted in J. Derrida, *Edmund Husserl's origin of geometry: An introduction.* Lincoln, NE: University of Nebraska Press.

Jablonka, E., & Klisinska, A. (2011). A note on the institutionalization of mathematical knowledge, or what was and is the fundamental theorem of calculus, really. In B. Sriraman (Ed.), *Crossroads in the history of mathematics and mathematics education* (pp. 3–40). Charlotte, NC: Information Age.

Kline, M. (1972). *Mathematical thought from ancient to modern times.* New York, NY: Oxford University Press.

Koudela, L. (2005). Curves in the history of mathematics: The late Renaissance. In J. Safrankova (Ed.), *WDS 2005. Proceedings of contributed papers, part I* (pp. 198–202). Prague, Czech Republic: Matfyzpress.

Laborde, C. (2005). The hidden role of diagrams in students' construction of meaning in geometry. In J. Kilpatrick, C. Hoyles, O. Skovsmose, & P. Valero (Eds.), *Meaning in mathematics education* (pp. 159–179). New York, NY: Springer.

Lakoff, G., & Núñez, R. (2000). *Where mathematics comes from. How the embodied mind brings mathematics into being.* New York, NY: Basic Books.

Otte, M., & Steinbring, H. (1977) Probleme der Begriffsentwicklung – zum Stetigkeitsbegriff [Problems of development of the concept—the concept of continuity]. *Didaktik der Mathematik, 5*(1), 16–25.

Presmeg, N. (2006). Semiotics and the "connections" standard: Significance of semiotics for teachers of mathematics. *Educational Studies in Mathematics, 61,* 163–182.

Radford, L. (2002). The seen, the spoken and the written: A semiotic approach to the problem of objectification of mathematical knowledge. *For the Learning of Mathematics, 22*(2), 14–23.

Radford, L. (2003). Gestures, speech, and the sprouting of signs: A semiotic-cultural approach to students' types of generalization. *Mathematical Thinking and Learning, 5*(1), 37–70.

Radford, L. (2006). The anthropology of meaning. *Educational Studies in Mathematics, 61,* 39–65.

Russell, B. (1901). Recent work on the principles of mathematics. *International Monthly, 4.*

Sardelis, D., & Valahas, T. (2012). The conics generated by the method of application of areas: A conceptual reconstruction. Retrieved from https://scirate.com/arxiv/1210.6842

Sfard, A. (2008). *Thinking as communicating: Human development, the growth of discourses, and mathematizing.* Cambridge, England: Cambridge University Press.

Sjöstedt, C-E., & Thörnqvist, S. (1963). *Analytisk geometri* [Analytical geometry]. Stockholm, Sweden: Natur och Kultur.

Tall, D., & Vinner, S. (1981). Concept image and concept definition in mathematics with particular reference to limits and continuity. *Educational Studies in Mathematics, 12,* 151–169.

Thom, R. (1973). Modern mathematics: Does it exist? In A. G. Howson (Ed.), *Developments in mathematical education: Proceedings of the Second International Congress on Mathematical Education* (pp. 194–209). London, England: Cambridge University Press.

Tietze, U-P (1994) Mathematical curricula and the underlying goals. In R. Biehler, R. W. Scholz, R. Sträßer, & B. Winkelmann (Eds.), *Didactics of mathematics as a scientific discipline* (pp. 41–53). Dordrecht, the Netherlands: Kluwer.

Christer Bergsten
Linköping University
Sweden

CHAPTER 3

祇園祭

Thoughts on Purification, Liminality, Art, Fermented Shark, Mathematics, and Education for Creativity

Paul Dowling

This chapter is about purification and liminality, and is presented in a stream of consciousness form. It was inspired by a paper by Eva Jablonka and Christer Bergsten (2010) in which they attempt what I shall describe as a purification of the category *theory* in mathematics education. My chapter consists of illustrative responses to a range of different kinds of texts. It will recruit analytic schemes from my organizational language, *social activity method* (SAM), and also will introduce new schemes relating to productive action and education for creativity; initially in dance, but then directed at other areas, including mathematics. The title of the chapter remains in its pure, Japanese form. Although (unlike the neologism that's coming up) easy enough to translate via the limination of the world wide web, but does it really matter? In this instance, this expression merely identifies the chapter. (I have, on second thought, appended a parenthetic subtitle to trap the search engines.)

Refractions of Mathematics Education, pages 49–75

[Beth:] Chat and the world chats with you, strive to address your issues and you'll be doing it alone (Kennedy, 2011).

It often seems to me that I'm talking to myself. Perhaps I should chat more. Well, here goes.

I knew exactly what I was doing on July 17 this year: I was sweltering in Kyoto's oven, really feeling for the traditionally clad marchers and tuggers in the parade of *yamaboko* floats that is the central event in the Gion Matsuri—the month long, Shinto festival that has been held in July pretty much every year since 970 AD (as far as I can ascertain, 天禄1 in the Japanese calendar). *Yamaboko* is a compound word that is made up of *yama* and *hoko* (with a necessary consonant change). These are the two different kinds of floats that are lugged around a 3-kilometer route by teams of up to 50 men for 2–3 hours. The 23 *yama* weigh 1.5 tonnes each, and the nine *hoko* up to 12 tonnes; they can be 25 meters high, with fixed wheels 2 meters in diameter, and have to be steered using wedges; float seems an almost singularly inappropriate term. There are lots of men crammed onto the multi-storied floats as well; all are wearing traditional *yukata* or other costumes, some of them are playing instruments (flutes and percussion), some of them are repeating simple movements with fans. The floats are very brightly decorated with tapestries and tassels (I have included some photos taken a year after this chapter was written, that is, on July 17, 2013; See Photos 3.1–3.5). At the front of the first *hoko* sits a young boy, who has been selected as a divine messenger. He is dressed in Shinto robes and a phoenix headdress, and he has been undergoing purification rituals for some weeks before the parade. In particular, he has been kept out of the presence of women. From 13th until the end of the parade—which he starts by cutting a rope—this boy has not been allowed to set foot on the ground. As I watch, the parade strikes me as peculiarly self-contained and self-referential; it attracts me as a spectator, but alienates me as any kind of participant in its meaning.

I am no longer really a tourist in Japan as I have lived there with my partner, Kimiko, for several months each year since 2000. I am not, though, really a Japanese resident either—not even a part-time one. I'm here on a tourist visa, and my fluency in the Japanese language is pathetic (Kimiko's excellent English is partly to blame for this). I am in a kind of liminal position: not tourist, not yet (part-time) resident. It's not just the Japanese language that remains elusive; events such as the Gion Matsuri parade remain alien to me. They seem to be about nothing but themselves; rather like the old men playing with toy airplanes in Kishine Koen park back in Yokohama. These are social occasions, of course, but their media seem to have no external referents. This is the view from my liminality. Strangely, it also seems to be the kind of view that many people have of sociology.

Photo 3.1 A Gion Matsuri Hoko.

Photo 3.2 Hoko Fans.

Photo 3.3　A Gion Matsuri Yama.

Photo 3.4　A Hoko Musical Section.

Photo 3.5   A Funeboko (a ship hoko: is it a boat or is it a hoko?)

Enough chat for the time being; I'll address some issues. If Beth is to be believed, this is where the readers skip to the next chapter in the book.

Gion Matsuri, Wikipedia advises me, "originated as part of a purification ritual (*goryo-e*) to appease the gods who were thought to cause fire, floods and earthquakes."[1] I've mentioned the ritual purification of the sacred boy, but there is more secular purification as well in the substantial exclusion of reference to contemporary cultural practice, though I did spot a couple—no more—contaminating watches, and spectacles that appeared to be immune from purifying removal. Boyd and Williams (n.d.) draw an alignment between purification in Shinto and formalism in art:

> The concept of purity in Shinto has three logical features. First, it establishes the distinction between the pure and the impure. Second, in the context of the tradition there is a difference in value between the two: purity is better than impurity. Third, the two contrasting states are related in a specific way. Compared to the pure, the impure has accretions or blemishes that are in principle removable; this is the relationship alluded to by the metaphor of the dust-covered mirror. In bare logical terms, there are two opposite, contrary notions or states, one of which is *in context* to be preferred to the other; and lastly, the lesser state can be viewed as blemished or as containing superfluous elements compared to the former.
>
> That the formal features of art share this same structure can be seen from what has already been said. Formalism describes a family of distinctions—

form vs. content, pattern vs. instance, or underlying structure vs. surface expression. (Boyd & Williams, n.d.)

The emphasized qualification, *in context* is important here; but I will return to that. I want to try to generalize the concept, *purification*, and I'll do this, firstly, by using a scheme that I've used before, and that I call a *practical strategic space*. The scheme is established via the Cartesian product of two variables. The strength of institutionalization (scaled I⁺/I⁻) refers to the extent to which a practice exhibits established regularity in the context, and at the level of analysis being considered. My own organizational language, SAM, for example, constitutes established regularity and is I⁺ within my own academic work and that of a number of users of the approach; but not at the level of analysis that represents sociology or educational research more generally, where it would be regarded as I⁻. The second dimension of my scheme is *discursive saturation* (scaled DS⁺/DS⁻), which refers to the extent to which the principles of a practice are rendered explicit within language. Again, SAM attempts to make its principles available within language (DS⁺) in a way that probably is not the case for the ritual fan movements of some of the participants in the Gion Matsuri parade, two of whom stand at the front of some of the *yamaboko*—at least, not in terms of the mechanism of their purifying action. The scheme is presented in Figure 3.1.

The purifying actions of the participants in the Gion Matsuri and those of Shinto priests described by Boyd and Williams, are constituted in the scheme as *skills*. These skills involve dressing and moving just so, and this possibly takes a lot of instruction and practice that may well not involve much in the way of explanation. Where such explanation becomes available, then this would constitute *discourse*.

I found another example of discursive purification on a short vacation in Iceland (a little more chat coming up; but perhaps you've already gone). I couldn't resist the challenge on the menu at a traditional Icelandic restaurant to try *kæstur hákarl*—fermented (rotten) shark —"if you dare!" The waiter explained to me, that this shark's flesh was poisonous if eaten straight out of the sea, because the fish had no urinary system and high levels of uric acid were concentrated in its body tissue. It needed to be purified, it seems. They achieved this by burying the shark in sand for three months

	Institutionalisation	
	I⁺	I⁻
DS⁺	*discourse*	*idiolect*
DS⁻	*skill* (competence)	*trick*

Figure 3.1 Practical strategic space (adapted from Dowling, 2009).

or so, and then hanging it up to dry for two or three weeks. This was not a magical practice, however: the mechanism of purification was quite clearly given and involved the expulsion of body fluids from the shark meat. This is *discourse* (DS^+/I^+). Unfortunately, this particular purifying action also had the effect of generating a substantial amount of ammonia in the fish. Mine was served in a sealable jar to minimize offense to other diners. I was to open the jar, take out a chunk of fish, stuff it quickly into my mouth and close the jar (and my mouth) again. The stench whilst the jar was open was appalling, but the meat tasted rather like blue cheese—perhaps that's the taste of ammonia, how would I know? (Actually, on my flight (first class) back from Tokyo, I discovered the flavor: camembert followed by a swig of nihon shuu—Japanese sake; this was an intellectually, though not particularly gastronomically, exciting experience.)

Before going further with the generalization of purification, I need to return to Boyd and Williams's description of the fundamental opposition constituted by formalism in art, and described as, "form vs. content, pattern vs. instance, or underlying structure vs. surface expression." They argue that formalism and Shinto rituals privilege the first term in each pair over the second, so that it is the recognition of that signified by the first term in that signified by the second; and in essence, the elimination of the excess or deviation that this implies that effects purification. Now I want to suggest (and I am, of course, far from being the first to do this) that these pairs do not helpfully indicate an ontological relationship. Rather, this might be understood as pointing to strategies that seek to establish alternative gazing discourses, and to privilege one via the use of signifiers that lend it a degree of synchronic and/or diachronic invariance. Looked at in this way, purification might go in either direction. Indeed, the longstanding opposition between structure and action in sociology and social theory works, in exactly the same way. My use (Dowling, 1998, 2009) of the device "(re)production" is an example of a defense strategy against (or a stimulation to) purifying strategies from either camp and a nod to, let's say, Bourdieu (1977), in *Outline of a Theory of Practice*, or Giddens (1984) in *The Constitution of Society*; or, most famously, Marx (1968[1852]) in *The Eighteenth Brumaire*. It is, though, surprising (at least, to me) just how much grip the (let's call it) formalist opposition—and the inevitable privileging of one side of it—has. Here is Stephen Hawking at the opening of the London 2012 Paralympic Games: "The universe is governed by rational laws that we can discover and understand" (from a note written shortly after the broadcast). It has all the philosophical sophistication of an episode of *Star Trek*. Just as an aside chat, *The Guardian* website today had a story asking the question, "Are bald men more powerful?"[2] The authors had presented paired images of famous men—including David Cameron, the current British Prime Minister—with and without hair and asked readers to vote. I'm not clear whether the vote went to underlying structure or surface expression.

I am not only following what we might refer to as the Derrida line—It is, of course, quite clear that structure and event (my preferred opposition) must entail one another—rather that their separation in acts of purification is always a strategy on behalf of a cultural arbitrary. Mathematics is one such cultural arbitrary, and structures—for example, those in formalist art and music—are often referred to in mathematical terms. The OECD, in a publication presenting sample PISA test items, defines the term "mathematisation" as involving:

- Starting with a problem in reality.
- Organizing it according to mathematical concepts and identifying the relevant mathematics.
- Gradually trimming away the reality to transform the real-world problem into a mathematical problem that faithfully represents the situation.
- Solving the mathematical problem.
- Making sense of the mathematical solution in terms of the real situation. (OECD, 2008, p. 99)

It is not entirely clear to me why the trimming away has to be done gradually. The interesting opposition here, however, is mathematics or reality. I (Dowling, 2010, 2013; Dowling & Burke, 2012) describe this kind of process in terms of *fetch* and *push* strategies. Fetching might be said to involve the mathematical purification (trimming away) of a non-mathematical practice. Pushing is not, however, the real world purification of a mathematical problem. That is because the subject of the action remains within the mathematical discourse, and purification must be actioned by the purifying subject.[3] Purification may, of course, be performed the other way around; which is to say, non-mathematical practice may legitimately recruit mathematical resources in its own elaboration, and then it is the mathematics that is trimmed away. The use of perspective grids in painting might be an example; you're not going to leave in the construction lines, are you? (Are you?)

Expression	Content	
	I⁺	I⁻
I⁺	*esoteric domain*	*descriptive domain*
I⁻	*expressive domain*	*public domain*

Figure 3.2 Domains of action scheme (from Dowling 2009).

Previously (Dowling, 1998, 2007, 2009, 2010, 2012, 2013), I introduced another schema (Figure 3.2) to describe what I am here calling purification. This distinguished the expression (signifiers) and content (signifieds) of an act or utterance in terms of their strength of institutionalization in the context in which the act or utterance was enacted. Here it is worth reflecting back on the use of this expression "in context" in the Boyd and Williams extract above; and again, purification can go in either direction once one admits that there are (at least) two discourses and not one, each one supposing itself to constitute the underlying structure of the other. Were this not to be the case, then a mathematical solution that is pushed back into the real world would always provide the optimal real world answer; in my experience, it generally does not. My scheme presents four domains of action. The *esoteric domain* refers to action that is strongly institutionalized within the relevant practice in respect to both expression and content; the *public domain* is action that is weakly institutionalized in respect of both expression and content; the *descriptive domain* refers to strongly institutionalized expression and weakly institutionalized content; and the expressive domain to weakly institutionalized expression and strongly institutionalized content. It is important to point out that the public domain is not the OECD's real world, but refers to any non-mathematical setting organized according to tacit mathematical principles. We might construe the expressive, descriptive and public domains as domains of liminality with respect to the purifying and to be purified discourses. This is to say that the public domain—and through it the expressive and descriptive domains—provide potential ways in to the esoteric domain. Potential, not necessary: in Dowling (1998) I demonstrated the way in which school mathematics texts largely confined students labelled as low ability to the public domain, and also that the principles of recognition of low and high ability resonated with low and high socioeconomic status (ses) markers, respectively. The outcome, in textual terms, was a state of permanent liminality for low ses mathematics students. It may be that this is a more general feature of selection in schooling across the curriculum. Boyd and Williams constitute liminality as a second feature of both Shinto ritual and formalist art. They describe liminality using a rather effective simile: "like the checker piece, temporarily lifted off the board in a different (vertical) dimension, while being moved from one square to another." One suspects that liminality is rarely a permanent state in Shinto, in art, or in checkers; although it seems to be so in the case of my position in Japan, as well as for low ses mathematics students.

The process of producing a doctoral thesis might be described as being conducted in a state of liminality (also seemingly a permanent state for some students). Here is some advice from Wendy Guthrie and Andy Lowe (2011) for aspiring classic grounded theory (CGT) researchers and their supervisors:

The only legitimate source of the classic approach to the GT research method is to be found in the publications of Dr. Barney Glaser and at the Grounded Theory Institute website (www.groundedtheory.com). Researchers using any other adaptation of GT will be deluding themselves and misleading others. The classic grounded theory research method is a very specific methodology with each step of the process very specifically delineated. Those who adapt and amend the process should not label the research method they have used as GT. Instead they should say their research was "influenced" or "inspired by GT" and then go on to create a new label for the research process they have used. (Guthrie & Lowe, 2011, p. 56)

Within the context of the edited collection (Martin & Gynnild, 2011) in which it appears, this purification strategy may legitimately be described as a skill. The collection institutionalizes, within itself at least, the identification of CGT as a specific method and warns against confusing it with modified GT approaches. Thus it purifies its esoteric domain. There is, however, no rationalizing of this position in the above extract, but it is constituted as discourse elsewhere. One of the specificities of CGT is that it attempts to access the key interests and concerns of the subjects in its research settings; it claims, in other words, a correspondence between its own public domain and the real world, rather like the OECD with respect to mathematics. For this reason, the CGT researcher cannot at the outset specify precisely what the research is about; hence the interdiction, in CGT, on the prior production of a literature review, which may end up being entirely irrelevant. At various points in the collection, it was also pointed out that the aversion to a prior literature review was also grounded on the need to avoid preconceptualizing the research setting, and ultimately to avoid imposing extant theory on the data that was being collected. McCallin, Nathaniel, and Andrews (2011), writing in the same collection, introduce morality into their purification strategy:

The moral foundation of discipline-specific research, turns on the ultimate goal of the profession. Classic grounded theory seeks truth in that its goal is to uncover important problems and patterns of social behavior as experienced, understood, and communicated by individuals absent of bias, value judgments, and interpretations of the theorist. It is a unique theory-generating approach to understanding human experience. (McCallin et al., 2011, pp. 78–79)

The exclusion of the "interpretations of the theorist" is not consistent with the position generally adopted in CGT and in this collection, which bases the theorist's interpretation on their theoretical sensitivity, which must be acquired through their engagement with theoretical work and by conducting previous CGT studies. This clearly entails differences between CGT researchers, and so differences in their interpretations. In effect, this establishes a distinction between public domain and the substantive settings. The particular purifying act by McCallin et al. that would eliminate interpretation, is therefore to be interpreted as an idiolect in terms of Figure 3.1. McCalin et al. elaborate this idiolect further:

> The moral imperative of research in the social sciences is to produce the best possible knowledge that can be used to positively affect those who require the services of a professional. So, there seems to be a valid moral justification for adherence to the tenets of classic grounded theory in disciplinary research. Furthermore, inadequate, skewed, misinformed, biased, or capriciously interpreted data and thoughtless, preconceived analysis of research data fails to attain the moral imperative central to disciplinary development.
>
> The suggestion here is that there are moral implications involved with remodelling of the original classic method. (McCallin et al., 2011, p. 79)

Purification here seems to eliminate all research that is not CGT, as immoral. Again, this is not consistent with CGT generally.

I am persuaded by much of the discourse of purification of CGT, not only by its presentation in Martin and Gynnild's anthology (Dowling, 2012), but also by the original seminal publication by Glaser and Strauss (1967); by its contrast with the subsequent work of Anselm Strauss working with Juliet Corbin (1990, 1998); and by Glaser's (1992) own response to the first edition of this work. It would seem from much that is reported—and left unreported—in Martin and Gynnild's book, that purification is not adequately achievable by discourse; it demands the acquisition of skill through mentoring—extended liminality—by a researcher who has previously been mentored by a qualified CGT researcher. This requirement effectively establishes a necessary pedigree of succession from Barney Glaser himself, the sole originator of true CGT. Furthermore, even "[l]earning the basics of grounded theory method takes about one and a half years of intensive work" (Gynnild, 2011a, p. 35), constituting what is, in effect, a priesthood. There is also an entry requirement for the seminary: "Grounded theory is only for people who are very intelligent and can conceptualize." Barney Glaser was speaking in an interview conducted and reported by Gynnild (2011b, p. 249). The association of intelligence with CGT, is a Glaser trick and so can be dispensed with. For the most part, however, a case is made for the specificity of the esoteric domain of CGT method and for the necessary purification both by skill and discourse.

Purifying further by eliminating the moral imperative of McCallin et al. allows for the legitimation of my own method—SAM—as an alternative to (not a competitor of) CGT. One of the specificities of SAM's esoteric domain, is the emergence of its analytic schemes, such as those in Figures 3.1 and 3.2, in transaction between the researcher and an empirical setting.

Such transaction is minimized in CGT, as it is the concerns of the research subjects that must dominate, hence the rejection of the preliminary literature review. SAM's transaction begins with a prejudicial statement of theoretical sensitivity: the sociocultural consists of strategic, autopoietic action directed at the formation, maintenance and destabilizing of alliances and oppositions, which are emergent upon the totality of such action and that are visible in terms of regularity of practice so that they are available for recruitment into subsequent action. CGT seeks to identify and conceptualize latent patterns in data (Simmons, 2011); SAM seeks to construct patterns in transaction between its own theoretical sensitivity and the research setting, which itself may be anything at all as is illustrated by the range of settings in this chapter. The result is the kind of conceptual scheme presented in this chapter, and an analysis of the setting in terms of these schemes. The former contributes to the *legacy* of previous analyses. SAM's theoretical sensitivity or *prejudice* constitutes an internal language (Bernstein, 2000; Dowling, 1998, 2009); the legacy constitutes an external language, external because it comes closest to the empirical field (Dowling & Brown, 2010), while the internal language is closer to the theoretical field.

I will continue just a little further with my discursive purification of SAM by eliminating two common criticisms that accuse schemes such as those in Figures 3.1 and 3.2 (and the others to follow) of undue reductionism; firstly on the grounds that continua or spectra would allow for greater delicacy in analysis. This accusation, however, is flawed in its presumption that it is possible to construct a continuum in the absence of a metric. For sure, it is possible to compare two texts or practices, and argue that one is more discursive than another that tends towards skill; or that one text exhibits a greater level of institutionalized action than another and, indeed, I have done just that above with reference to chapters in the CGT collection. This, however, is achieved only by identifying the presence or absence of discourse and skill or I^+ and I^- strategies in each text. It is necessary to move to a level of analysis at which a single strategy can be identified, and then to move back up to the level of the whole text or practice ,to describe the accumulation and distribution of strategies. This, in fact, does enable the production of a metric, and I have conducted quantitative analyses of various school mathematics texts in Dowling (1998, 2009, and 2013). In Dowling (2013), for example, I have analyzed items in Foundation and Higher Tier GCSE papers, and have shown that esoteric and expressive domain strategies are about twice as frequent as public and descriptive domain strategies. These pairings of strategies are apposite, because the esoteric and expressive domain strategies both signify I^+ mathematics, while public and descriptive domain strategies both signify I^- practice; or, in this case, non-mathematical practice. The second criticism accuses the schemes of totalizing texts or practices. As I have just explained and demonstrated, this is simply untrue.

It is true that the pedagogic recontextualising of SAM generally requires the provision of exemplars from its own public and descriptive domains, which is to say, text or practices that has been analyzed by SAM. A short cut is to identify a whole practice, such as physics, as discourse or pottery as skill, in order to draw on the stereotypical images that the audience is likely to possess to introduce the category DS. In a live presentation, should a member of the audience object that pottery is other than this, and that it in fact draws routinely on substantial discursive theory having to do with clay and glaze composition, drying times and firing temperatures, not to mention the historical and aesthetic discourses of ceramics; then this will serve admirably to enhance the introduction to and further illustrate the use value of the concept and the method more generally.

In the preamble, I mentioned that this chapter had been inspired by a paper by Eva Jablonka and Christer Bergsten (2010). In this paper, they set out to purify (my term) the category 'theory' in mathematics education research. They identified two categories, *intertextuality* and *relational density*, both scaled high/low and defined as follows:

> As with all research, mathematics education is discursive in nature and can only be understood in reference to previous research. However, the intertextuality can be more or less explicit (as for example by use of specialised language, references to intellectual roots, building on previous research outcomes). In the examples we discussed above, another dimension emerged, that is, the extent to which relations between the key concepts are established. We refer to this dimension as relational density. (Jablonka & Bergsten, 2010, p. 37)

This generates another scheme that is reproduced in Figure 3.3. My initial response was to consider what kinds of strategies were constituted in different practices, starting, oddly, with the Gion Matsuri. I know it's not research, doesn't claim to be theory, and it certainly is not mathematics education; but I have a tendency to want to put apparently unconnected things together to see what happens. More of this later. As a pure spectator, the parade presented itself to me very much as an ad-hoc construction; although if we relax the requirement—I may be reading this requirement

| | Relational Density | |
Intertextuality	High	Low
High	*theory*	*conglomerate*
Low	*local model*	*ad-hoc construction*

Figure 3.3 Different modes of classifying, modelling or theorising. (Jablonka & Bergsten, 2010)

into Jablonka and Bergsten inappropriately anyway—that relational density and intertextuality be discursive, then perhaps it's more of a local model. It seemed, from my perspective, connected within itself, but quite disconnected from anything else.

I think it's what I took to be the discursive imperative in the scheme that stimulated an itch, when I began to think about it. I am fully in accord with Jablonka and Bergsten's attempt to upgrade research—I've long felt dissatisfaction with the kind of pretensions they're aiming at—but perhaps not quite for the same reasons. For example, I'm not sure that research can be understood only in relation to former research. Certainly this seems to be a requirement of some academic journal reviewers. One anonymous reviewer of a paper that I recently submitted, wanted me to engage with a substantial body of research—possibly including their own—that they deemed relevant. This would have taken about a year's work, lengthen an already lengthy paper considerably, and would not, as far as I could see, enhance the arguments. Furthermore, if this reviewer had identified the connections on their own, why did they want me to do it as well? Other readers having different theoretical sensitivities, would presumably have found other connections; and had the revised paper gone to a different reader with demands similar in form but different in content, the paper might have ended up in a permanent state of liminality. The reviewer also pointed out, that they had published in both mathematics education and sociology, and had also studied methodology; the latter was a remark in response to my mention of grounded theory. This lead to a demand for further discussion and more literature. In general, this reviewer seemed to feel that, although they could understand the work well enough, no one else would be able to. Consisting primarily of assertions, this review seemed to be a skilled attempt—skilled in the sense that it adopted fairly standard reviewing strategies and made reference to published work—at purification of the field. The strategy did not work, however, as two other reviewers were very positive about the paper, and didn't want much changed (just a clarification of the abstract, in fact) and the editor agreed with them. An explicit engagement with previous research is a presentation of the way that the author has responded to and recruited that work. It is an important strategy in relation to the formation, maintenance, and destabilizing of alliances and oppositions within the relevant academic field; how many of us go directly to the bibliography of a paper to see if our name or the names of any of our friends or enemies are there? The literature review section of a paper also constitutes discursive purification, in respect of explicating its originality and positioning it in the field. This appears to be the principal function of the *post hoc* literature review in CGT. The review might also, perhaps, help to clarify some of the concepts in a paper by reference to already familiar concepts. This latter point is perhaps close to what Jablonka and Bergsten were getting at, but I don't see this as a necessary feature. Quite often one suspects that items are inserted into a paper in an attempt to persuade the reader that the author has read and understood something that the reader probably found too difficult (Deleuze & Guattari, 1984; Lacan, 1977).

Relational density is a quality that is certainly to be valued in research, but again, I am not convinced that it is essential, particularly if relationality must be expressed discursively. I note, for example, that my own and other institutions in the United Kingdom are permitting practice-based research to be submitted at the doctoral level. Practice-based research is explained by Linda Candy thus:

> Practice-based research is an original investigation undertaken in order to gain new knowledge partly by means of practice and the outcomes of that practice. In a doctoral thesis, claims of originality and contribution to knowledge may be demonstrated through creative outcomes in the form of designs, music, digital media, performances and exhibitions. Whilst the significance and context of the claims are described in words, a full understanding can only be obtained with direct reference to the outcomes. (Candy, 2006, p. 1)

Significance and context are discursive; but central to the act of reading and understanding the research, is the experience of the creative artifact or performance that may well not be. The relationality within the Gion Matsuri parade is well established by a high density of quotes in dress and design, by movements, and so forth. Intertextuality may also be achieved in the same kind of way without necessarily moving to language. None of this is to say that language is unimportant—that would be crass—but connections may be established in both DS⁺ and DS⁻ modes as is widely demonstrated in the arts.

As I have said, I am entirely in accord with the attempt to upgrade research, and the use of Jablonka and Bergsten's scheme as an interrogator of research will certainly encourage this appropriately. Furthermore, I don't object to the attempt to purify the term theory. I do not refer to SAM as a theory, but as a method, like CGT, despite the presence of the word theory in the name of the latter: theory is what CGT generates—discovers in its original presentation (Glaser & Strauss, 1967)—not what it is. I think my general objection is to work that is unethical by virtue of a lack of investment on the part of the author.

Another scheme is about to be born. For my other dimension, I will refer to SAM's fundamental internal language that is concerned with strategic action directed at the formation, maintenance, and destabilizing of alliances and oppositions. A strategy directed at the formation of an alliance is tantamount to an attempt to destabilize another; and, similarly, destabilizing an opposition is an attempt to form a new alliance. So productive action can

	Production	
Investment	Maintaining	Destablising
Low	*chat*	*accident*
High	*craft*	*art*

Figure 3.4 Productive action.

be described as maintaining or destabilizing production involving high or low investment, which gives rise to the scheme in Figure 3.4.

So my ethical objection is to chat or accident being paraded as research. This does not mean that they are always unwelcome or unproductive, indeed, I am referring to all modes as productive—here in relation to the maintenance or destabilizing of alliances and oppositions. The original discovery of penicillin was apparently largely accidental, although its realization as a useable drug was certainly a combination of craft—the deployment of existing knowledge in biochemistry and pharmacology—and art, including creative juxtapositions. My ethical objection is to the fact that the name that is attached to the drug in popular memory is Alexander Fleming, who had the accident; and the names of Howard Walter Florey, Ernst Chain, and Norman Heatley, who did most of the creative work, are often ignored (Heatley didn't even get a Nobel Prize). Chat is, as Beth implies, social glue.

The scheme in Figure 3.4 is not in competition with Jablonka and Bergsten, but is complementary; it provides another way of thinking about and analyzing productive action. I do wonder, though, where they would locate this and others of my schemes in theirs. Thinking particularly about *Art*—as opposed to art —it is probably easy to come up with examples of chat and craft, and the elephant art gallery (http://www.elephantartgallery.com) has examples of what I might call accident: paintings not of, but by elephants. In February of this year, I attended an art exhibition: "Context" at the Michaelis Galleries, University of Cape Town. The UCT website described the focus of the exhibition as follows:

> Context draws together artists who use the book-object as a conceptual point of departure for the exploration of the printed text. The artists_projects engage the history, value and institutional importance afforded to the book-object. The works on display grapple with the materiality and influence of the idea of the book and the way the notion of the book is related to artistic practice. (http://www.michaelis.uct.ac.za/newsevents/exhibitions?viewExhibition=172)

The exhibition was curated by Fabian Saptouw and opened by the American artist, Mark Dion. In his opening speech, Dion declared that "[A]rt should challenge." This, to me, signaled an interest in *Art* as art, rather than (or at least in addition to) *Art* as craft (I'm not sure what Dion would think of low invest-ment *Art*). In other words, *Art* should destabilize cosy alliances. Well, let's see.

Saptouw's own presence in the exhibition included two works that I want to refer to here. The first was titled *The Picture of Dorian Grey*. This work consists of two sheets of paper that can be viewed at http://www.iart.co.za/all-images-archive/open-books/saptouw1_resize.jpg/ and http://www.iart.co.za/all-images-archive/open-books/saptouw2_resize.jpg/. It may be viewed on the Internet, and the description given on the gallery label was:

Mondi Rotatrim 80 g/m² Paper, Photocopier Toner—TN311, entire text of Oscar Wilde's novel The Picture of Dorian Grey; 210×297 and 210×297 mm.

For those reading this without Internet access (or in case the links have rotted), the first page of this work seems to have been achieved by photo-copying the pages of the novel on top of each other on a single sheet, result-ing in a black rectangle, with some striated variation in depth of blackness and ragged edges. The second sheet seems to be a count of the number of occurrences of each typewriter character (a, b, c, . . . ^p, /, —) within the novel. The second work was *Of Grammatology* (I have not yet been able to find any images on the Internet.[4]) Its label announced:

Mondi Rotatrim 80 g/m² Paper, Photocopier Toner—TN311, entire text of Jacques Derrida's Of Grammatology; Dimensions variable.

This piece consisted of a stack of pages containing the text, but in alpha-betical order: all the a's, then all the b's then all the c's, all the way down to all the hyphens (I think) with each new character beginning a new page. Accompanying this was a page with another character count.

When we came out of the exhibition, my colleague lamented that she couldn't really engage with this kind of work; she lacked the appropriate ex-periential field and so could not recognize the quotes. This seemed to echo my own response to the Gion Matsuri parade; feeling anomie in the face of the text, though not—at least, not in my case—repelled by it. In each case, the arcane construction of the text, effectively purified its legitimate audi-ence. Now it may be that the artists who exhibited—and even those respon-sible for the parade—had, in some respect or another, deployed pedagogic strategies. This kind of strategy involves the author of a text attempting to retain control over the principles of interpretation of the text (Dowling, 2009). Here, it might be the artist had distinct messages that they intended to convey, and one of the mechanisms available to them was intertextual quotations. In order to get the message, the audience would have to rec-ognize these; those failing to do so by virtue of their lack of appropriate experiential field, would be purified out of the pedagogy.

On the other hand, it has often been suggested, that any response to a work of *Art* is legitimate. This constitutes *Art* in terms of what I refer to as exchange strategies; the author seeks to hand control over the principles of interpretation to the audience (Dowling, 2009). This is *Art* refusing to purify itself in respect to its audience; any experiential field will do. So, reflecting on Saptouw's two works I wondered:

- About the experience of the artist in producing the works and the experience of artistic production generally, how does it compare with, say, my experience in writing sociology;
- Why *The Picture of Dorian Grey* and *Of Grammatology?* These are quotes, of course, that will possibly differentially purify the audience.
- Is there a resonance between the accumulation of ink on the page and the aging of the picture and, ultimately, of Dorian Grey himself? What kind of images does this generate in terms of ageing generally?
- Is there a resonance/dissonance between deconstruction in Derrida and the dismantling of his work in Saptouw's? In a sense, the now widespread use of deconstruction when dismantling would be more appropriate, and constitutes, perhaps, a naïve quotation when contrasted with what would appear to be Saptouw's ironic one.
- Would it change the work if I removed a page from the stack? If so, would it change only for me—unless discovered—or for everyone?
- Does it matter whether or not the artist miscounted or made a typographical error? How about such errors—numerical, typographical, spelling, grammar, errors in argument—in this chapter (there are bound to be many)?
- Saptouw (as both artist and curator) has followed standard practice in labeling his work, but at what point does this become ironic?

I could go on, and each of these thoughts might birth an essay, albeit an artistically and philosophically ill-informed one; I am, after all, only a sociologist. The point, however, is that any experiential field will do (and I'm now thinking of a song from *Joseph and the Technicolour Dreamcoat*—sorry, so much chat when what I should be doing is art and craft). If, however, I imagine Saptouw's work as an exchange text, I deprive him of control over its interpretation. So how does *Art* challenge? What is the nature and limit of its pedagogic authority, if it is deprived of interpretive control; what is the degree zero, the maximal purification of *Art* as presented by Dion?

My answer: in order to challenge, *Art* as an exchange strategy can only draw attention to itself within, of course, the context of the *Art* presentation (after all, a road traffic accident draws attention to itself in other contexts). Saptouw's work certainly succeeds here. Drawing attention must in some respect be destabilizing, and destabilizing must be construed as the opposite of purification; which is to say, contamination. Saptouw contaminates by his recruitment of works from another field (although *Art* usually does this), and by his seemingly arbitrary and pointless dismantling and reconstruction of it; this stands in stark contrast with the craft strategies that are prevalent in popular imaginings of *Art*, in the *Art* that most commonly gets to be exhibited by the Royal Academy or on biscuit tins, postcards, and populist pottery. All of this purified art has been done already; where does "Art must draw attention to itself" take us?

It takes me, firstly, to Marcel Duchamp's readymades and to *Bottle Rack* (1914),[5] in particular, which seems to claim that the ultimate purification of *Art* is the artist's choice. Then it takes me to Piero Manzoni's *Socle du Monde* (1961);[6] again, the artist has the whole world to choose from, and chooses the whole world. But, for me, the ultimate (so far) is Rachel Whiteread's *Untitled Monument*,[7] which occupied the unoccupied plinth in Trafalgar Square for a short period in 2001 and 2002 and, in particular, on an August afternoon, resplendent in its sunshine halo (my photograph demurely hiding amongst thousands of others in a cupboard). Whiteread's work has included casts of living spaces, first a room—*Ghost* (1990)[8]—and then a whole house—*House* (1993);[9] the latter criminally demolished (not deconstructed, surely) by the local authority. Her corpus also includes the commisioned Judenplatz Holocaust Memorial (2000) in Vienna.[10] *Untitled Monument* takes the approach to a new level. This is how I described it in *Sociology as Method*:

> Whiteread's sculpture is a clear resin cast of the fourth plinth in Trafalgar Square, London, which it occupied between June 2001 and May 2002. This plinth has until recently been unoccupied, the others celebrating famous generals and King George IV on a horse (a statue commissioned by the king himself). Whiteread's cast was placed inverted on top of the plinth recalling, perhaps, Manzoni's inverted plinth. Forty years on from Manzoni and with the benefit of Baudrillard's insight, it is now possible to read this work as a questioning of the artist. It inverts Manzoni's plinth (technically I suppose it is not Manzoni's plinth that is upside down, but the viewers), as the mirror image of the plinth on which it stands and which precedes it as the condition of existence of Whiteread's work qua art. The transparency of the work also reveals the physical condition of existence of the plinth itself, which is the space that it consumes.[11] Whiteread's work signals a system of monumentalizing practices that always precede the monument and that simulate it as production, rather than as merely reproductive of the practice of monumentalizing. (Dowling, 2009, p. 51)

All of this, and it appeared on that sunny August afternoon to be truly beautiful, rendering shameful the graffiti on the concrete plinth, "What for?" (Chat or accident? It depends which side of the square you're on.)

The invoking of Baudrillard's (1993) structural law of value—invoked by my responding to Whiteread's work in terms of exchange strategies—now raises the question: what are the processes whereby a work comes to be exhibited. This question goes sociologically beyond what is the meaning of the work in and of itself, either as a pedagogic or as an exchange strategy. It also casts me back to Saptouw's work. What are the structures that it might be calling into question? *The Picture of Dorian Grey* purifies an interpretation of the production process involved in writing or reading a book, as an overwriting of one page by the next, metaphorically, a life as a succession of rewritings of its history, the total recall of which obscures all specificity. *Of Grammatology* purifies the signifiers to their lowest level of analysis; the structural law of value always operates at multiple levels; purification to degree zero eliminates rather than enables meaning, but immediately allows it back in at a meta-level. Both works render their respective texts unreadable, as they were (presumably) authored, disabling pedagogy and any security in comprehension. I refer to the arbitrariness of signification, to the arbitrariness of meaning making, to the arbitrariness of reading: any dream will do. Self-referentiality *par excellence*: Gion Matsuri.

Liminality must (in one way or another) come to an end, particularly as I sense the gaze of my word limit (though no matter: there will be no extended concluding remarks to this chapter, it's strictly stream of consciousness). So, take your partners for the last dance—dance education, in fact. Last year (it's 2012 now) an MA student, Monica Bernardo (2011), came to me wanting to study for her dissertation the ways in which students and teachers understand creativity in dance. The topic had recently become particularly significant, as teachers were now being asked to assess creativity in students' performances. In reflecting on her own teaching of dance and in interviews with students and teachers, she discovered that students would deploy various strategies in producing their project performances when they were asked to work creatively. One approach was to mix genres of dance taken from the dance curriculum: traditional Indian dance with tap, for example. Another was to introduce dance moves from outside of the dance curriculum. These might include moves that the students had seen on TV (today, wouldn't it just have to be *Gangnam Style*). Other students copied moves from outside of dance altogether: the actions of sports players or of a kitten playing with a ball of wool. One student said that she revised science by getting her mother to question her from the textbook, while she (the student) bounced on a trampoline. We didn't ascertain whether this rather unique approach worked, but the student did recall that it had occurred to her that she might incorporate trampoline-like movements into her dance project. Bernardo also discovered that it was the students who were most confident in dance, who borrowed from beyond the curriculum; and students who were the least confident, who tended to remain within it. You may have

Activity Specificity	Institutionalization	
	I⁺	I⁻
Internal	*curricular*	*popular*
External	*alternative curricular*	*free*

Figure 3.5 Modes of creative action.

Activity Specificity	Institutionalization	
	I⁺	I⁻
Internal	*dance curricular*	*popular dance*
External	*sports moves*	*kitten/trampoline*

Figure 3.6 Modes of creative dance.

guessed as much, but purification of this discourse reveals another analytical scheme; it's shown as Figure 3.5.

I can now populate the scheme with the illustrations from Monica's research; I've done this in Figure 3.6.

This needs some explanation. Selecting moves from the dance curriculum is a comparatively safe strategy; the moves are legitimized as dance, and as dance institutionalized by the school. Moving from institutionalized dance to popular dance is a somewhat riskier strategy, because legitimization liminally awaits assessment. It would appear to the student, perhaps, that the dance curriculum purifies what can count as dance. Borrowing from sports action, returns institutionalization insofar as the particular moves borrowed are recognizeable: cricketers' bowling action, for example. This is again risky, however, because it is an import from what might be described as an alternative curriculum. Most risky, however, is the introduction of kittenish and informal trampolining actions, because they are neither legitimate dance—inside or outside of the school—nor are they institutionalized moves. The cline in risk, curricular, popular or alternative curricular; free did seem to coincide with what Bernardo told me about her assessments of the confidence in dance of the students: the more confident the student, the riskier—in terms of my scheme—the strategy.

Before considering the relevance of this scheme for education, it is worth considering its generalizability to other parts of the curriculum. In order to do this, it is essential to keep in mind that I have considered only the movement element of dance, and not (for example) music or costume. If I am to focus on another school subject, I will have to purify these in the same way, and be explicit about what aspect of the subject I am going to focus on.

Activity Specificity	Institutionalization	
	I⁺	I⁻
Internal	*English curriculum*	*fan fiction*
External	*historiography/narrative research*	*enacted narrative*

Figure 3.7   Modes of creative writing.

Let's take creative writing and concentrate on the construction of narrative. Figure 3.7 proposes appropriate examples for the creative action scheme.

Fan fiction involves the production of narratives relating to extant literary or media texts (Chung, Dowling, and Whiteman, 2004). Recruiting narrative ideas from this popular form of creative writing, or from forms of I⁺ narrative construction that are external to literature as such would (generalizing from Figures 3.5 and 3.6) entail a higher level of risk than remaining within the English curriculum. Enacted narrative would result from a narratively purified action in any context; introducing such forms into school-based creative writing, would involve maximum risk. Figure 3.8 presents some examples for creative action in mathematics. Here I am focusing on the study of formal systems. It is notable that I have been unable to come up with an example of a popular form of mathematics. At a recent seminar that I held at the University of Cape Town, it was suggested that ethnomathematics might qualify. This, however, is not mathematics in the same sense that popular dance is classified as dance and fan fiction is classified as creative writing. Indeed, ethnomathematics is most definitely not classified as mathematics by participants in the relevant practice; and this is understood as a cause for concern on the part of some authors on the subject. I take a different view: see the discussion on ethnomathematics and the myth of emancipation in Dowling (1998). In any event, ethnomathematics identified in the construction of fishing baskets (for example), is tantamount to suggesting that the cricket bowler is actually dancing or that the historiographer (in contrast with, say, Hilary Mantel) is engaged in creative writing;[12] the terms are being appended metaphorically, not literally. Ethnomathematics, then, draws together pretty much every activity that

Activity Specificity	Institutionalization	
	I⁺	I⁻
Internal	*curricular mathematical systems*	*no popular mathematics*
External	*games, management, governance,*	*natural systems*
	sociology, ethnomathematics	

Figure 3.8   Modes of creative mathematics.

is I⁺ in some relevant context, and that can be described in mathematical terms. Natural systems are not socially institutionalized, and I have not yet heard a mathematics educator claim that the Fibonacci numbers that can be associated with the growth pattern of a celery plant are the result of the celery doing mathematics, and that celery emancipation will ensue should this knowledge be revealed to it.[13] Of course, the distinction between social and natural institutionalization is also an analytical act—the primal act that calls the social into existence—and not an ontological one. Unless, that is, we want to deny that humanity is a natural phenomenon.

The scheme in Figure 3.5 and its interpretation in terms of risk and creativity, offers an interpretation of the category, creativity, and a suggestion in respect to education for creativity. The interpretation—and I am not at all claiming originality here—is that a creative act may involve putting together elements that have previously been kept apart: synthesis. This is contamination. Its opposite is the separating of that which has previously been kept together—analysis—which is purification. Both are concerned with the maintenance or destabilizing of alliance and opposition. Creative analysis, however, also involves the imposition of a principle that had hitherto not been incorporated into the relevant practice; so also it involves contamination, though at a different level of analysis. I am going to say, then, that creativity necessarily involves contamination in one form or another and the scheme in Figure 3.5 presents a way of looking at this in terms of increasing risk in the face of an assessment regime. Some students will adopt high-risk strategies right from the start, and if the principle that they are deploying in their contaminating act is legitimate—a contamination of sources of movement in dance or narrative in creative writing or formal systems in mathematics—then their creativity may be well received. Others will be initially reluctant to take risks, and may be encouraged by the use of the sequence proposed here: curricular, popular/alternative curricular, free. My recollection of being taught dance in primary school, was that it started with being told to pretend to be a teapot, a freeing and high-risk strategy that made some of us feel rather stupid. I'm sure that dance education has moved on in the many years since then.

I have promised no extended concluding remarks, but I won't just stop here. I have just (last night) finished reading Sam Thompson's first novel, *Communion Town: A city in ten chapters* (2012). This work was longlisted (but disappointingly for me and presumably for Thompson, not shortlisted) for the Man Booker Prize. Here is the end of James Purdon's review in *The Observer*:

> The danger with this kind of pastiche is that it can become a form of literary homeopathy, diluting its source to the point where nothing of value remains. While *Communion Town* sometimes seems willing to acknowledge that dan-

ger, it never quite succeeds in overcoming it. (http://www.guardian.co.uk/books/2012/sep/09/communion-town-sam-thompson-review)

I just couldn't disagree more, though suspect that Purdon would write an even harsher review of this chapter. Robert-Douglas Fairhurst, writing in *The Telegraph* and awarding the work five stars out of five, seems to me to have been more willing to pay attention to the book:

> Perhaps it isn't surprising that one of his best stories involves a boy who constructs a model town on the floor of his sitting room, allowing his imagination to stretch out and discover what it can do. He could be a figure of the young novelist at work. Turning the pages of Communion Town you become aware that here is a new writer working out what he can do, and realising that he can do anything. (http://www.telegraph.co.uk/culture/books/bookreviews/9378707/Communion-Town-a-City-in-Ten-Chapters-by-Sam-Thompson-review.html)

When I first saw Robert Altman's *Short Cuts*, it occurred to me that I could have constructed a unity of the various narratives—a penchant of mine at the time—by attaching a particular identity to a corpse that appeared at the end of the film. As it turned out, this proved impossible (as I recall, by virtue of gender), and the unity unravelled on the screen. I have learned, since then, to read film and novels without such purifying tendencies; to yield to the text without establishing it in the public domain of a sclerotic wisdom. That way, they last, as art or indeed as craft, far longer. The Gion Matsuri parade is now a summer away, but its contaminating juxtaposition with Eva and Christer's scheme (Figure 3.3) has given me a wonderful fortnight autumn vacation. Now back to the marking.

NOTES

1. http://en.wikipedia.org/wiki/Gion_Matsuri
2. http://www.guardian.co.uk/science/poll/2012/oct/04/bald-men-more-powerful, article by Patrick Barkham, photomanipulation by David McCoy.
3. Elsewhere (for example, in Dowling, 1998, 2007, 2010, 2013; Dowling & Burke, 2012) I have preferred the term *recontextualisation*, which is less directional than purification; but the latter term, I think, has more grip in the context of the present chapter.
4. I have subsequently found an image which, though unlabelled, appears to be the *Of Grammatology* stack of pages as one of the sequence at http://www.parking-gallery.net/2012/05/05/context-at-the-michaelis-galleries-cape-town/
5. See the image at http://www.toutfait.com/unmaking_the_museum/Bottle%20rack.html.

6. See the image at http://radicalart.info/everything/Manzoni/socle_du_monde-s.jpg
7. See the image at http://www.artbabyart.com/braveworld/Whiteread.htm
8. See the image at http://1.bp.blogspot.com/-5XoxXGVM9sI/TbS8bvZ4uRI/AAAAAAAAAAM/KyEc269-kS8/s1600/a000575e.jpg
9. See the image at http://2.bp.blogspot.com/-VquOcukdmWU/UGB98_YR-PsI/AAAAAAAABvE/l_0NZtwBQ0Q/s1600/whiteread-house.jpg
10. See the image at http://www.tripadvisor.com.au/LocationPhotoDirectLink-g190454-i64523895-Vienna.html
11. And so marks out this mode of monumentalizing from the other that is prevalent in Trafalgar Square (and elsewhere in Westminster) that is eloquently spoken about by Mark Wallinger's piece, *Ecce Homo*, that was the first work to appear on the empty plinth.
12. Simon Schama (1992) plays with the tension here. His book includes a dedication to John Clive, "for whom history was literature". It might be argued that Hilary Mantel (2012) does the same from the opposite camp.
13. But see the mathematically interesting discussion at http://www.branta.connectfree.co.uk/fibonacci.htm and the bizzare assertion at http://www.bbc.co.uk/news/health-22991838; we could yet see a stick of celery appointed professor of mathematics.

REFERENCES

Baudrillard, J. (1993). *Symbolic exchange and death.* London, England: Sage.

Bernardo, M. (2011). Encouraging creativity in dance composition tasks and lessons. (Unpublished master's thesis). Institute of Education, University of London.

Bernstein, B. (2000). *Pedagogy, symbolic control and identity.* New York, NY: Rowman & Littlefield.

Bourdieu, P. (1977). *Outline of a theory of practice.* Cambridge, MA: Cambridge University Press.

Boyd, J. W., & Williams, R. G. (nd.). Shinto purification rituals: An aesthetic interpretation. Retrieved from http://sunsite.berkeley.edu/jhti/shinto/project1.html

Candy, L. (2006). *Practice based research: A guide.* Sydney, Australia: Creative & Cognition Studios, University of Technology.

Chung, S-y., Dowling, P. C., & Whiteman, N. (2004). (Dis)possessing literacy and literature: Gourmandising in Gibsonbarlowville. In A. J. Brown & N. Davis (Eds.), *The world yearbook of education 2004: Digital technology, communities and education.* London, England: Routledge.

Deleuze, G., & Guattari, F. (1984). *Anti-Oedipus: Capitalism and schizophrenia.* London, England: Athlone.

Dowling, P. C. (1998). *The sociology of mathematics education: Mathematical myths/pedagogic texts.* London, England: Falmer Press.

Dowling, P. C. (2007). Quixote's science: Public heresy/private apostasy. In B. Atweh, A. C. Barton, M. C. Borba, N. Gough, C. Keitel, C. Vistro-Yu, & R. Vithal

(Eds.), *Internationalisation and globalisation in mathematics and science education* (pp. 173–198). Dordrecht, the Netherlands: Springer.

Dowling, P. C. (2009). *Sociology as method: Departures from the forensics of culture, text and knowledge.* Rotterdam, the Netherlands: Sense.

Dowling, P. C. (2010). Abandoning mathematics and hard labour in schools: A new sociology of knowledge and curriculum reform. In C. Bergsten, E. Jablonka, & T. Wedege (Eds.), *Mathematics and mathematics education: Cultural and social dimensions. Proceedings of MADIF7* (pp. 1–30). Linköping, Sweden: SMDF.

Dowling, P. C. (2012). Being Barney Glaser. *Grounded Theory Review, 11*(2).

Dowling, P. C. (2013). Social activity method: A fractal language for mathematics. *Mathematics Education Research Journal, 25,* 317–340. Retrieved from http://link.springer.com/article/10.1007%2Fs13394-013-0073-8

Dowling, P. C., & Brown, A. J. (2010). *Doing research/Reading research: Re-interrogating education.* London, England: Routledge.

Dowling, P. C. & Burke, J. (2012). Shall we do politics or learn some maths today? Representing and interrogating social inequality. In H. Forgasz & F. Rivera (Eds.), *Towards equity in mathematics education: Gender, culture, and diversity* (pp. 87–104). Heidelberg, Germany: Springer.

Giddens, A. (1984). *The constitution of society: Outline of the theory of structuration.* Cambridge, England: Polity.

Glaser, B. G. (1992). *Basics of Grounded Theory analysis: Emergence versus forcing.* Mill Valley, CA: Sociology Press.

Glaser, B. G., & Strauss, L. (1967). *The discovery of grounded theory: Strategies for qualitative research.* Piscataway, NJ: Aldine Transaction.

Guthrie, W. & Lowe, A. (2011). Getting through the PhD process using GT: A supervisor-researcher perspective. In V. Martin & A. Gynnild (Eds.), *Grounded theory: The philosophy, method, and work of Barney Glaser* (pp. 51–68). Boca Raton, FL: BrownWalker Press.

Gynnild, A. (2011a). Atmosphering for conceptual discovery. In V. Martin & A. Gynnild (Eds.), *Grounded theory: The philosophy, method, and work of Barney Glaser* (pp. 31–49). Boca Raton, FL: BrownWalker Press.

Gynnild, A. (2011b). Living the ideas: A biographical interview with Barney G. Glaser. In V. Martin & A. Gynnild (Eds). *Grounded theory: The philosophy, method, and work of Barney Glaser* (pp. 237–253). Boca Raton, FL: BrownWalker Press.

Jablonka, E., & Bergsten, C. (2010). Theorising in mathematics education resesarch: Differences in modes and quality. *Nordic Studies in Mathematics Education, 15*(1), 25–52.

Kennedy, A. L. (2011). *The blue book.* London, England: Jonathan Cape.

Lacan, J. (1977). *Écrits.* New York, NY: W. W. Norton.

Mantel, H. (2012). *Bring up the bodies.* London, England: Fourth Estate.

Martin, V., & Gynnild, A. (Eds.). *Grounded theory: The philosophy, method, and work of Barney Glaser.* Boca Raton, FL: BrownWalker Press.

Marx, K. (1968). The eighteenth brumaire of Louis Bonaparte. *Selected works in one volume. K. Marx and F. Engels.* London, England: Lawrence & Wishart.

OECD. (2008). *Taking the test: Sample questions from OECD's PISA assessments.* Paris, France: OECD.

McCallin, A., Nathaniel, A., & Andrews, T. (2011). Learning methodology minus mentorship. In V. Martin & A. Gynnild (Eds.), *Grounded theory: The philosophy, method, and work of Barney Glaser* (pp. 69–84). Boca Raton, FL: BrownWalker Press.

Schama, S. (1992). *Dead certainties: Unwarranted speculations.* New York, NY: Vintage.

Simmons, O. E. (2011). Why classic grounded theory. In V. Martin & A. Gynnild (Eds.), *Grounded theory: The philosophy, method, and work of Barney Glaser* (pp. 15–30). Boca Raton, FL: BrownWalker Press.

Strauss, A., & Corbin, J. (1990). *Basics of qualitative research: Grounded theory procedures and techniques.* London, England: Sage.

Strauss, A., & Corbin, J. (1998). *Basics of qualitative research: Second edition: Techniques for developing grounded theory.* London, England: Sage.

Thompson, S. (2012). *Communion town: A city in ten chapters.* London, England: Fourth Estate.

Paul Dowling
The Institute of Education, London
United Kingdom

CHAPTER 4

MUSINGS ABOUT MODELS AND MODELING IN MATHEMATICS

Michael N. Fried

As one who was a serious student of applied mathematics at a certain point in his mathematical life, and who worked on mathematical models for biological phenomena, I should know what a model is. Yet, I confess I have never fully persuaded myself that I really do. It is no wonder then that mathematical models and modeling were the subject of more than one conversation that I have had with Eva Jablonka. This was not only because of Jablonka's longstanding interest in modeling in mathematics education, beginning with her doctoral dissertation; but also because of a philosophical temperament that allowed her to take seriously a simple question of the form (What is that, really?). After good conversations such as those with Jablonka, one was always left going over the arguments; thinking about what might have been added, or left out, a good example, a better example. In short, one continues to muse about the subject. It seems fit in this *liber amicorum* that I should give back to her, as a true offering of friendship, some of these musings on mathematical models; not a report of research results or a survey of current literature, but only the loose ends of a conversation with a thoughtful friend.

Refractions of Mathematics Education, pages 77–85
Copyright © 2015 by Information Age Publishing
All rights of reproduction in any form reserved.

The problem of what a mathematical model is, has only been compounded by the interest the subject has generated in mathematics education since about the 1980s; an interest that, for some, has situated mathematical modeling at the very center of mathematics education. This is pointedly clear in the case of PISA, since its emphasis on mathematical modeling flows ineluctably from its definition of mathematical literacy; that is, its definition of what it is to have been mathematically educated to begin with. The current PISA definition is as follows:

> Mathematical literacy is an individual's capacity to formulate, employ, and interpret mathematics in a variety of contexts. It includes reasoning mathematically and using mathematical concepts, procedures, facts, and tools to describe, explain, and predict phenomena. It assists individuals to recognise the role that mathematics plays in the world and to make the well-founded judgments and decisions needed by constructive, engaged and reflective citizens. (OECD, 2013, p. 25)

PISA then makes mathematical literacy a reflection of one's ability to relate mathematical procedures, strategies, and concepts to a not-necessarily-mathematical world. In practice, this means beginning with some phenomenon not obviously mathematical in character, and endowing it with mathematical structure in order to gain insight, to make predictions, or control the phenomenon itself. This is mathematical modeling precisely in the sense most familiar to people. It is what people generally mean when they refer to mathematical models for biological systems, for the weather, for social interactions, or for economic policy.

One can understand the virtue of connecting mathematics to the world beyond mathematics, especially the world in which one works and acts, and the world of which one is a citizen. But there are also theoretical difficulties arising immediately from this. To start, there is the difficulty of defining what constitutes this world. Is it a purely natural world, objective and politically neutral? Or is it a world that is constituted by social, institutional, cultural, or political forces, even when it appears neutral? The history of science during the last half century has made it clear that the latter must in fact be taken into account. The implications of this for mathematics education and mathematics curricula has been of particular interest toJablonka (Jablonka, 2007). In my other role as an historian of mathematics, this is also of interest to me. However, here I would like to put these social and cultural issues aside, and consider the problem of modeling within rather traditional categories of mathematics.

Doubtlessly, the architects of PISA know that there is more to mathematics than these areas of applied mathematics. Yet by placing them at the heart of mathematical literacy, one has the impression that these (and therefore modeling) are at the heart of mathematics itself. This is one way, then, that the problem of understanding what mathematical modeling is, becomes compounded by the mere emphasis placed on it: for how does modeling relate, if at all, to *pure* mathematics? Or put more trenchantly, does the central focus on mathematical modeling exclude pure mathematics as a part of mathematics education, except, maybe, as a set of tools for applications? Is one justified studying number theory, without its application to cryptography?

A slightly different aspect of the centrality of modeling gives a partial answer to this question. This is the tendency to see mathematical modeling as a kind of model itself for problem-solving. In other words, instead of being only an example of problem-solving or a certain kind of mathematical activity involving skills of problem solving, modeling contains the essence of problem solving altogether. Thus, Lesh and Lehrer (2003) have described modeling in completely general terms:

> The term *models* here refers to purposeful mathematical descriptions of situations, embedded within particular systems of practice that feature an epistemology of model fit and revision. That is, "modeling" is a process of developing representational descriptions for specific purposes in specific situations. (p. 109)

And English and Sriraman, in their reflections on problem solving for the 21st century (English and Sriraman, 2010) have argued explicitly that, "A rich alternative to [the other approaches for teaching problem-solving] is one that treats problem solving as integral to the development of an understanding of any given mathematical concept or process. Mathematical modeling is one such approach" (p. 271). With this kind of general position on mathematical modeling, the idea of a model and the process of modeling must be taken as equally appropriate for pure mathematics and applied mathematics, just as problem solving itself has always been.

This centrality and generality of mathematical modeling may not be wrong. It may also be true that mathematical modeling addresses the understanding of pure mathematical concepts as maintained by English and Sriraman. However, another view of the matter, and another justification, is gained by recalling that although models, as I have said, typically call to mind *applied* mathematics, the truth is that models belong quite as much to *pure* mathematics as they do to applied. Mere cognizance of this fact answers the difficulty of the seeming exclusion of pure mathematics from model-centric mathematics education. And considering models in both contexts, sheds light on what a model is, though not completely.

So let me go back to the beginning and ask again: what is a model. I shall begin with a model as it is understood in pure mathematics.

A *model*, for the pure mathematician is quite simply an interpretation or a realization of a formal system of axioms. To illustrate, consider a system[1] containing three types of objects, *P*'s, *Q*'s, and *R*'s, governed by the following axioms:

1. A *Q* is a set of *P*'s, containing at least two *P*'s.
2. Two distinct *P*'s are contained in one and only one *Q*.
3. An *R* is a set of *P*'s, containing at least three *P*'s which do not belong to the same *Q*.
4. Three distinct *P*'s which do not belong to the same *Q* are contained in one and only one *R*.
5. If an *R* contains two distinct *P*'s in a *Q*, it contains all the *P*'s in the *Q*.
6. If two *R*'s have a *P* in common, then they have a second *P* in common.

As it stands, this is truly a *formal* system: the *P*'s, *Q*'s, and *R*'s have *no content*; they are mere blanks to be filled in. To give the system content, one associates with the *P*'s, *Q*'s and *R*'s real concrete objects that are related to one another in a way consistent with the axioms. The result is a *model* for the system. For example, let the *P*'s be the signs a, b, c, the *Q*'s the pairs {a, b}, {b, c}, {a, c}, and *R* the one set {a, b, c}. By means of the simple diagram in Figure 4.1, where the *P*'s are the vertices of the triangle, the *Q*'s the sides, and *R* the whole triangle, one has no trouble seeing that the axioms 1–6 are all satisfied:

Another model takes triples of real numbers (x, y, z) as *P*'s: triples satisfying two linear equations, ax + by + cz + d = 0 and ex + fy + gz + h = 0, as *Q*'s; and triples satisfying one linear equation ax + by + cz + d = 0 as *R*'s. In yet another model, ordinary points in Euclidean 3-space are the *P*'s, lines the *Q*'s, and planes the *R*'s. Here too it is not hard to verify that the axioms are satisfied. The importance of the model in this context, is that it provides an *image* by which the logical aspects of the formal system can easily be seen. For instance, if the *trivial* first model is consistent, as it most certainly is, then so is the formal system itself; and, hence, is any model for it, *trivial* and *non trivial* alike (see Rosenbloom, 1950, p. 15). In other words, the consistency of the trivial three-point geometry, proves the consistency of all Euclidean geometry with regard to its incidence relations. Let us now turn to models in the first context: namely that of applied mathematics.

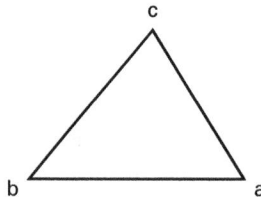

Figure 4.1

I own a book called *Introduction to Probability Models* (Ross, 1972). The name is curious, for upon opening this book, one will find a book on probability very much like any other. Yet the name is right. Every application of probability is really the making of a probability model. I shall explain by means of an example. Suppose one is watching a bird on a wire: the bird moves back and forth on the wire, now a few steps right, now a few steps left. One wants to know such things as whether the bird will ever reach the end of the wire, or how often the bird will come back to its starting position. Now one knows, of course, that the bird is a complex creature in a complex environment, and that the *reasons* why the bird moves right or left depends on all sorts of factors from the turbulent motion of the air, to the complicated patterns of electrical discharges in the bird's neurons. To gain some insight into the movement of the bird along the wire, one creates an image. One considers the bird as a featureless point moving on a line, right and left; in steps equal in length and equal in time. One assumes, moreover, that the point moves right or left with a certain measurable likelihood: say r for a right step and l for a left step. These likelihoods are themselves a realization of a certain axiomatic system; namely, that governing probability measures. *Within this image*, one can now answer some questions. For example, in n steps the bird (the point) will return to its starting point with a probability:

$$p = \begin{cases} \begin{pmatrix} n \\ \dfrac{n}{2} \end{pmatrix} r^{\frac{n}{2}} l^{\frac{n}{2}} & n \text{ even} \\[12pt] 0 & n \text{ odd} \end{cases}$$

This is so because to return to its starting point, the bird must (in any order) make an equal number of right steps and an equal number of left steps, so that n must be even: the number of right and left steps must be $n/2$. We have thus gained some information about the movement of the bird, at least to the extent that our image is a good one; and in this connection Akira Okubo has said:

One must not forget that analogies are no better than analogies, models nothing more than models, and hypotheses simply hypotheses. The mere fact that a mathematical model agrees well with a small amount of data does not suffice, insofar as the agreement could be coincidental. (Okubo, 1980, p. 4)

On the other hand, in one of the many conversations I had with Okubo, he said to me that the most essential thing always was to have a clear picture in your mind.

The model in applied mathematics is a mathematical image of the phenomenon one is interested in studying; and in working with a mathematical model, one is really working with the image rather than the phenomenon itself. The bird's movements are quite determined, but we look at them *as if* they are by chance.

What is particularly fascinating about models in applied mathematics, is that quite different phenomena can have the same mathematical image: the same model. In this way, the different phenomena can be viewed as images of one another; one might say models of one another via the *mathematical* model. For example, suppose we have an animal population that grows with a certain growth rate. If the growth rate were constant, then we would say that the change in the total number of animals is proportional to the number of animals present at a given time. This is modeled then by the equation $\frac{dN}{dt} = RN$ where $N(t)$ is the number of animals at time t, and R is the constant rate of growth.[2] But it is more reasonable to assume that the growth rate itself depends on the number of animals present: as the population reaches an equilibrium level K owing to a limited amount of living space or food supply, the growth rate becomes 0. But the further the population is below K, the more positive will be the growth rate; and if it exceeds K, it should be negative. The growth rate *itself* we can model as a linear relation, $R = r(1 - \frac{N}{K})$, which we note approaches r as N approaches 0; is 0 when $N = K$; and is negative for $N > K$. The model then for our population is the equation:

$$\frac{dN}{dt} = r\left(1 - \frac{N}{K}\right)N$$

where we assume N takes some given value at time 0: say $N(0) = N_0$. Now, the point is this. We can consider a very different phenomenon, the spread of an innovation. Considering how people communicate with one another, and how a population can become saturated with information (just as an environment can become saturated with animals), we arrive at an identical equation; *now* N represents the number of people at a given time t who know about the new innovation (Braun, 1978). Thus, via the single mathematical model (the image of the phenomena in the equation), we can think of the growing animal population; in turn, as a model for the growing knowledge of a new innovation; or the growing knowledge of an innovation as a model for a growing animal population.

On first sight, it appears that the idea of a model in pure mathematics and that in applied mathematics are opposites: the pure mathematical model being a concrete interpretation of a formal axiomatic system, and the applied mathematical model being an abstraction of a concrete phenomenon. But a closer look shows that there is really no contradiction. In both cases the final model is an interpretation of some set of formal axioms;

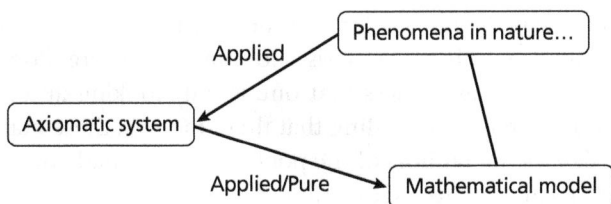

Figure 4.2

only in the case of pure mathematics do the axioms serve as a starting point, whereas in applied mathematics they serve as mediators between the non-mathematical phenomena and the mathematical interpretation. The diagram in Figure 4.2 illustrates the relationship.

More importantly, there is a play of images in both cases. But perhaps it is more suggestive to use a literary comparison and call a model a *mathematical metaphor*.[3] We call a bird a point moving randomly on a line, or by a set of formal rules, a geometry: these are true metaphors, and I mean that in the usual *literary* way.[4] In literature, as well as in plain speaking, one does not move very far without encountering a metaphor; in fact, metaphors are so ubiquitous one all too often forgets one is using a metaphor at all. This situation was the subject of a beautiful essay by C. S. Lewis. In that essay, Lewis said,

> When we pass beyond pointing to individual sensible objects, when we begin to think of causes, relations, of mental states or acts, we become incurably metaphorical. We apprehend none of these things except through metaphor: we know of the ships only what the *Kennigar* will tell us. Our only choice is to use the metaphors and thus to think something, though less than we could wish; or else to be driven by unrecognized metaphors and to think nothing at all. I myself would prefer to embrace the former choice, as far as my ignorance and laziness allow me. (Lewis, 1969, p. 146)

The examples of models given in this note were chosen not only because they well illustrate the way in which models in mathematics are like metaphors in literature, but also because they are, like the metaphors that one uses without realizing one is being metaphorical; models that one might not immediately recognize as models. Indeed, even young students encounter models in mathematics far more than they are led to believe. In what is known as the *new program* in Israel (even though it is hardly "new" anymore), for example, students learn that vectors can be considered as ordered n-tuples (or directed line segments, arrows): both are only *models* for the axioms of a vector space. Without that understanding, students are often left asking, sometimes in deep frustration, "So what *is* a vector: an n-tuple or an arrow?" One *needs* these models, these images, these meta-phors "to think something," as Lewis said; but somewhere along the line, one must develop an awareness that one is only looking at a model. No less important is the understanding that these models, these mathematical metaphors, like the metaphors of the poet, are things made by us: we make them *so that* we can see more clearly and deeply.

NOTES

1. This is an adaptation of the axioms of "incidence geometry" as they appear in Prenowitz and Jordan (1965).
2. That the model may be given by an equation makes sense of a phrase that I at least as a student of applied found confusing at first, namely, "to solve a model."
3. I have recently discovered that a similar comparison has been made in the writings of the mathematician Rutherford Aris (1994).
4. Metaphors in the specific sense of Lakoff and Nunez (2000) have been used before in connection with models (Williams and Wake, 2007), but I wish to keep metaphors in their literary setting. This is to emphasize that in connection to models they are not only a way of seeing something but also the result of a deliberate and creative act of the imagination.

REFERENCES

Aris, R. (1994). *Mathematical modelling techniques.* New York, NY: Dover.
Braun, M. (1978). The spread of technological innovations. In M. Braun, C. S. Cole-man, & D. A. Drew (Eds.), *Differential equation models* (pp. 91–97). New York, NY: Springer-Verlag.
English, L., & Sriraman, B. (2010). Problem solving for the 21st century. In B. Srira-man & L. English (Eds.), *Theories of mathematics education: Seeking new frontiers* (pp. 263–290). Heidelberg, Germany: Springer.
Jablonka, E. (2007). The relevance of modelling and applications: Relevant to whom and for what purpose? In W. Blum, P. Galbraith, H-W. Henn, & M. Niss (Eds.), *Modelling and applications in mathematics education: The 14th ICMI Study* (pp. 193–200). Berlin, Germany: Springer.
Lakoff, G., & Nunez, R. E. (2000). *Where mathematics comes from: How the embodied mind brings mathematics into being.* New York, NY: Basic Books.
Lesh, R., & Lehrer, R. (2003). Models and modeling perspectives on the develop-ment of students and teachers. *Mathematical Thinking and Learning, 5*(2 & 3), 109–129.
Lewis, C. S. (1969). Bluespels and flanansferes. In A. M. Eastman (Ed.), *The Norton reader* (pp. 136–148). New York, NY: W.W. Norton.

OECD (2013). "PISA 2012 assessment and analytical framework: Mathematics, reading, science, problem solving and financial literacy." OECD Publishing. Retrieved from http://dx.doi.org/10.1787/9789264190511-en

Okubo, A. (1980). *Diffusion and ecological problems: Mathematical models.* Berlin, Germany: Springer-Verlag.

Prenowitz, W., & Jordan, M. (1965). *Basic concepts of geometry.* New York, NY: John Wiley.

Rosenbloom, P. (1950). *The elements of mathematical logic.* New York, NY: Dover.

Ross, S. M. (1972). *Introduction to probability models.* New York, NY: Academic Press.

Williams, J., & Wake, G. (2007). Metaphors and models in translation between college and workplace mathematics. *Educational Studies in Mathematics, 64,* 345–371.

Michael Fried
Ben Gurion University of the Negev, Beer Sheva
Israel

CHAPTER 5

MATHEMATICS AND THE YELLOWING OF IDEOLOGIES

Uwe Gellert

I recently had the chance to inherit a modest collection of textbooks. Curiosity took over, and I found myself combing through the volumes and running over the pages. Most of the books had been published in Germany from 1850 to 1950; some exceptions came from France, the United Kingdom, and the United States. I was tempted to start a comprehensive historical and sociological textbook analysis, à la Dowling (1998), looking for strong and weak classification of signifiers and signified, and how this all was related to the political context of the time. Needless to say, I soon realized that this would easily convert into a substantial research program. Being aware of the limits of my intellectual perseverance, I resigned from that self-administered task.

What I am going to present here, is a pale reflection of what those textbooks could offer to the serious analyst. With nothing but a good dose of superficiality, I would like to draw attention to some photos, drawings, and other illustrations. My reading of the illustrations is piecemeal and incomplete; the selection idiosyncratic and driven by oddities. It is meant as a "non-contribution" in a sense.

In the first part of this chapter, the topic will be the practical value of mathematics. I will focus on the contexts from which a child's mathematics might arise, and the application of mathematics displayed in the textbooks.

Refractions of Mathematics Education, pages 87–98
Copyright © 2015 by Information Age Publishing

I will argue that these contexts relate to the seriously conflictive, political situation of the time.

The second part of the chapter will be dedicated to the soul of the learner. Is mathematics for all? How can teachers deal with the different psychological dispositions of their students? We will glance at a textbook for prospective mathematics teachers in educational psychology, and will discuss a classification of mathematics learners based on the Wartegg drawing completion test.

THE "LIFE SURROUNDINGS"
FOR THE STUDY OF MATHEMATICS

Mathematics, as has been demonstrated (Davis & Hersh, 1981), plays an important role in the lives of citizens of technologized societies, where "technologized" includes the formalization of societal and social relationships. Technology in this respect, is the anchor point for the conceptual duality of mathematization/demathematization (Gellert & Jablonka, 2007). In these textbooks, we found a selection of those aspects of citizens' lives, and the selection was not made randomly, but was due to a more or less overt political ideology.

Figure 5.1 was page 2 of *Büttners Berliner Rechenbuch*, published by Oskar Porthe in 1937; it included a commentary for the teacher saying: "Correspondent to the nature of the child, the numbers are deduced from

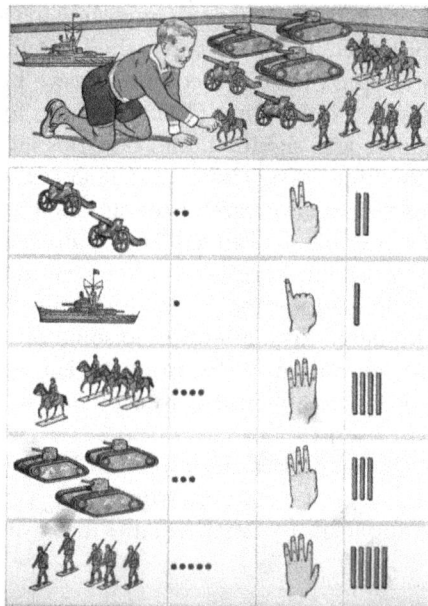

Figure 5.1 *Büttners Berliner Rechenbuch*, page 2.

Figure 5.2 *Büttners Berliner Rechenbuch,* page 3.

infantine life surroundings, here the numbers 1 to 5 from playing with soldiers." The boy's life and surroundings reflected a militarization of German society at that time; this context was the framing of the early numbers activity. It is a tragedy that students of the respective age group were forced into so called Hitler's "letztes Aufgebot" [last contingent] at the end of World War II. Against this background, the prospective life surroundings of girls (Figure 5.2) seemed harmless, although firmly integrated into German National Socialist ideology.

Arguably, there seems to be empirical evidence that boys rather than girls play with military toys (today, too), and that girls rather than boys play with dolls. A mathematics textbook's purpose is necessarily pedagogic, but there is no necessity to reinforce and normalize the dichotomy of damage and care.

Weapons of destruction as a road to mathematics are not a monopoly of German educators. Figure 5.3 shows the book cover of a survey of

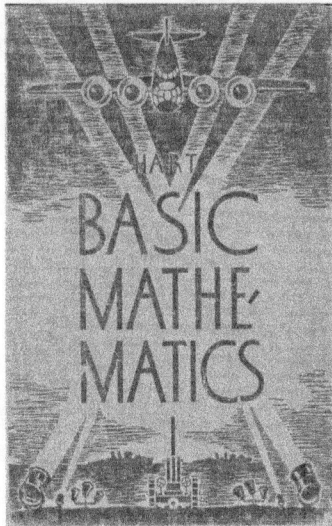

Figure 5.3 *Basic Mathematics,* cover.

secondary mathematics for study of pure and applied science, including engineering, particularly in the context of industrial production and the armed forces (Hart, 1942).

The book cover's drawing was illustrated with a photo of an aircraft bomber, the "flying fortress" (Figure 5.4), and other military aircrafts were used as illustrations within the book (Figure 5.5).

Similar schemes were used for mathematics exercises (Figure 5.6). The survey seems not only to transmit mathematical knowledge for its intended readership, but also the fields of application. By this way, students were learning mathematics together with the ideology behind its use.

Figure 5.4   *Basic Mathematics,* frontpiece.

Figure 5.5   *Basic Mathematics,* page 144.

Aircraft Spotters' Guide, National Aeronautics Council, Inc.

CURTISS P-40

13. *Quotients in real problems are usually inexact.*

Illustration. The wing area of the Curtiss P–40, known as the Tomahawk in the R.A.F., is 236 sq. ft.; its span is 37.33 ft. What is the mean chord (or the average width) of its wings?

The quotient is 6.32 ft. Since 236 has three significant *digits*, we'll keep 6.32 ft. as the result.

To get a result correct to tenths, the quotient must be carried out to hundredths.

$$
\begin{array}{r}
6.32 \\
37.33)\overline{236.00\ 00} \\
223\ 98 \\
\hline
12\ 02\ 0 \\
11\ 19\ 9 \\
\hline
82\ 10 \\
74\ 66 \\
\hline
\end{array}
$$

Figure 5.6 *Basic Mathematics*, page 16.

It seems to be an interesting field of study to scrutinize the contexts in which mathematical application exercises are formulated (Jablonka, 1996). Apparently, the well-known sequence "nationalism, imperialism, war" had been a grounding structure of mathematics applications for the learning of mathematics, even if the ideological intentions of the textbooks were not as explicit as they had been in the former illustrations. Consider, for instance, a book with the slightly sardonic title, *Elementary Algebra: Without Answers* (Smith, 1877), which, as the title might promise, was mainly an intra-mathematical treatise. But there were some examples of the power of algebra (p. 295f.; although nowadays the classification of the mathematical domain could be different); and we might have a look at the contents:

- Out of 100 soldiers how many different parties of 4 can be chosen?
- There are 12 soldiers and 16 sailors. How many different parties of 6 can be made, each party consisting of 3 soldiers and 3 sailors?
- Out of 12 consonants and 5 vowels how many words can be formed, each containing 6 consonants and 3 vowels?

Figure 5.7   *Gut und richtig Rechnen,* cover.

The last example seemed to offer a rather abstract concept of "word"; probably the concept of "soldier" should be regarded in a similar way. As an indication of the reasonableness of this assumption, consider that the book, different from what seems to have been common at the time, did not specialize its readership. Apparently it was written for the reader interested in esoteric mathematics.

Other textbooks were more explicit in their targeted readers (see Figure 5.7). It translates roughly as: "Computing good and correctly, or: a small school of calculations for self study. Containing all elementary mathematics, with complete solutions for the given examples. Popularised for functionaries, merchants, servicemen, and individuals." The examples, however, did not explicitly draw on the readers' fields of occupation. This was remarkably different in a comprehensive book on mercantile arithmetic (Feller & Odermann, 1866). For a reader unfamiliar with the mercantile context, the mathematics exercises would have seemed virtually incomprehensible (see Figure 5.8).

A difference in the popularized book for functionaries, merchants, servicemen and individuals was also visible in the book cover (Figure 5.9) of a seventh-grade mathematics textbook (Baßler et al., 1949).

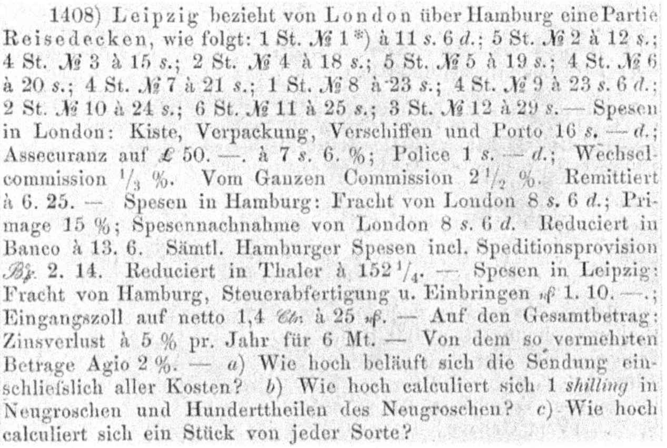

1408) Leipzig bezieht von London über Hamburg eine Partie Reisedecken, wie folgt: 1 St. № 1 *) à 11 s. 6 d.; 5 St. № 2 à 12 s.; 4 St. № 3 à 15 s.; 2 St. № 4 à 18 s.; 5 St. № 5 à 19 s.; 4 St. № 6 à 20 s.; 4 St. № 7 à 21 s.; 1 St. № 8 à 23 s.; 4 St. № 9 à 23 s. 6 d.; 2 St. № 10 à 24 s.; 6 St. № 11 à 25 s.; 3 St. № 12 à 29 s. — Spesen in London: Kiste, Verpackung, Verschiffen und Porto 16 s. — d.; Assecuranz auf £ 50. —. à 7 s. 6. %; Police 1 s. — d.; Wechselcommission ⅓ %. Vom Ganzen Commission 2 ½ %. Remittiert à 6. 25. — Spesen in Hamburg: Fracht von London 8 s. 6 d.; Primage 15 %; Spesennachnahme von London 8 s. 6 d. Reduciert in Banco à 13. 6. Sämtl. Hamburger Spesen incl. Speditionsprovision ℬℨ 2. 14. Reduciert in Thaler à 152 ¼. — Spesen in Leipzig: Fracht von Hamburg, Steuerabfertigung u. Einbringen ₰ 1. 10. —.; Eingangszoll auf netto 1,4 ℔ à 25 ₰. — Auf den Gesamtbetrag: Zinsverlust à 5 % pr. Jahr für 6 Mt. — Von dem so vermehrten Betrage Agio 2 %. — a) Wie hoch beläuft sich die Sendung einschließlich aller Kosten? b) Wie hoch calculirt sich 1 *shilling* in Neugroschen und Hunderttheilen des Neugroschen? c) Wie hoch calculirt sich ein Stück von jeder Sorte?

Figure 5.8 *Das Ganze der kaufmännischen Arithmetik,* page 513.

Figure 5.9 *Unser Rechenbuch,* cover.

Here mathematics seems to have been everywhere, although relevant mainly for manual work. (The textbook had been accredited for school use by the Education and Religious Affaire Branch, Office of Military Government [United States]. Was the Morgenthau Plan lurking there?) The occupations involving mathematics were small-scale, and there were many examples in this textbook that applied mathematics to the many situations modestly sketched on the book cover. Some years before, in contrast, the textbook drawings (Figure 5.10 and Figure 5.11) had been rather pretentious (Lötzbeyer & Molthan, 1929).

To sum up, mathematics textbooks offered the reader a wealth of plainly visible connections to mathematics in its domains of application. Dominant ideologies at the time, are apparent from the temporal distance of the contemporary reader. It might be assumed that future generations will have

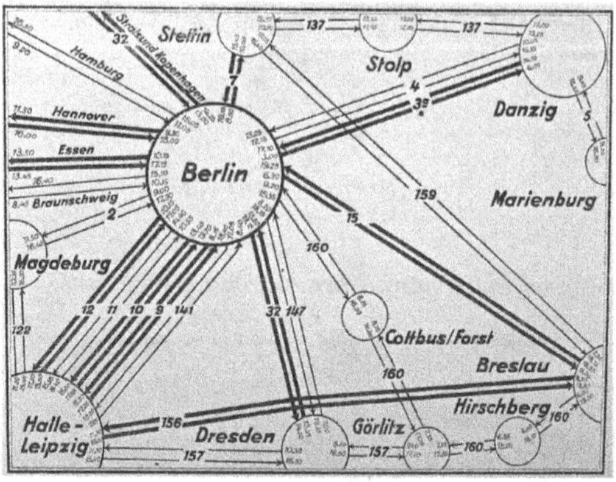

Figure 5.10 Rechenbuch für höhere Schulen; Taken from the route network of German Luft-Hansa.

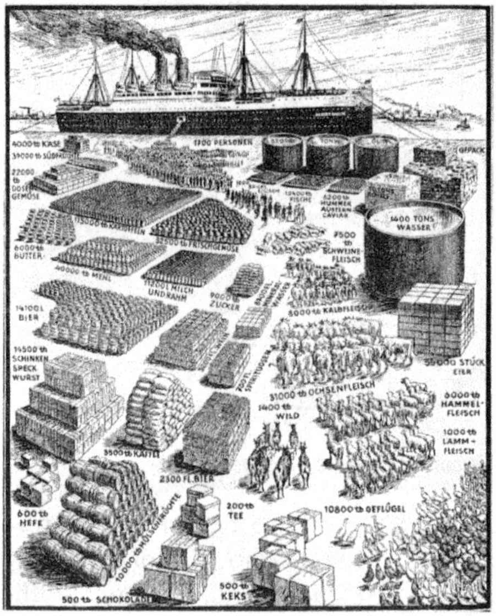

Figure 5.11 Rechenbuch für höhere Schulen; Resolve problems from the drawing "The Victualing of the Steamer Ballin"

no difficulty identifying the hidden ideologies present in the mathematics textbooks of today. I suspect, consumerism and a certain fascinations for technological artifacts will be good candidates then.

MATHEMATICS AND THE SOUL OF THE LEARNER

A different kind of textbook was a book used for instructing future mathematics teachers. The volume *Pädagogische Psychologie des mathematischen Denkens* (trans. *Educational Psychology of Mathematical Thinking*), was published in 1953 by Kurt Strunz; it sets the scene in words of wisdom which, in their particular prose, could probably be appreciated only by a German-speaking reader: "Die Seele des Zöglings, das Bildungsgut und die Persönlichkeit des Lehrers sind die drei Pole, in deren Kraftfeld sich alles erzieherische Tun der Schule abspielt" [The soul of the young, the educational material and the personality of the teacher are the three poles in their force field (in which) is happening all educational activity of the school]. To draw on the student, the teacher, and on the subject matter was, of course, no singularity. Interestingly, the book offered a typology of students' psyche, and the psyche's positioning toward mathematics. There was, first, the extrovert passive sentimentalist: this person was classified as the "anti-mathematician." Second, the empirical thinker, the scientist, the practitioner; who had a more clinical perception and an inclination towards objectivity, at least to a larger extent than the extrovert sentimentalist. Third was the introvert sentimentalist; this person could be an anti-mathematician or a worshipper of mathematics. Fourth was the conceptual theorist whose characteristics resulted in a positive disposition toward mathematics.

More interestingly still, the book offered a test for indexing students within this classification: it was the Wartegg drawing completion test (Wartegg Zeichen Test), which was developed by the psychologist Wartegg (1939). The test form was DIN A4 and included eight squares on a black background in which small signs were given (Figure 5.12).

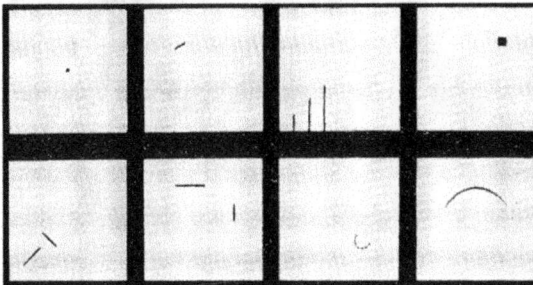

Figure 5.12 Wartegg Zeichen test.

As illustrative examples for the use of the Wartegg Zeichen Test for the classification of students, four completed tests of 13- to 15-year-olds were presented (Figure 5.13 to 5.16). The classification labels were given in the captions.

Figure 5.13   Solutions of a musically gifted eidetic.

Figure 5.14   Solutions of an extrovert sentimentalist.

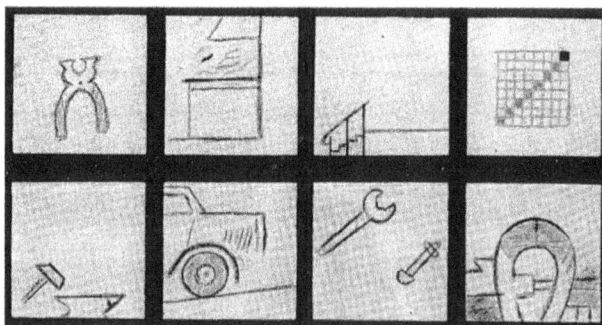

Figure 5.15   Solutions of an extroverted, manually gifted, down-to-earth, self-reliant boy.

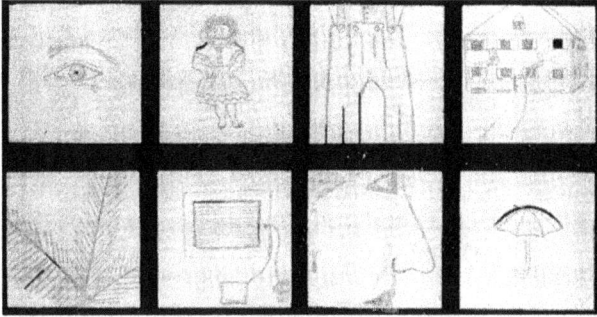

Figure 5.16 Solutions of a mathematically gifted, determined student.

In the appendix, the book provided a guideline for the character analysis of completed test drawings. Unfortunately, specialist studies in psychology were called upon as a requirement for the successful work with the Wartegg Zeichen Test.

The character classification was the sorting device for a differential administration of mathematical knowledge. The manually gifted and the mathematically gifted would "naturally" be provided with a different kind of mathematics education. But, as Strunz (1953) holds, the educational aim was to shape all student characters not only according to their "proper mental lifestyle" (p. 161), but also to the non-corresponding types of perception and thinking:

> Erstens Emporbildung eines jeden Zöglings zu dem ihm angemessenen geistigen Lebensstil, z.B. zum mathematischen Praktiker, zum Theoretiker, zum Wahrheitssucher, zum Bewunderer mathematischer Schönheiten, zum Denker mit Esprit oder zum gründlichen Denker; zweitens aber auch Hinführung des Jugendlichen zum Verständnis der ihm nicht entsprechenden Grundformen des Auffassens und Denkens und, soweit es möglich ist, zur geistigen Umstellung auf diese (p. 161f.). [trans. First development education of every young to him reasonable intellectual lifestyle, E.g., to the mathematical practitioners, the theoretician for the truth seeker, to the admirers of mathematical beauty, to the thinkers with ESPRIT and the thorough thinker; Secondly but also education of the young people to understand of the not corresponding basic forms of setting up and thinking and, as far as it is possible on the mental transition to this.]

By these lofty words, educational ideology and scientific pretension were coalesced.

REFERENCES

Baßler, E., Bäurle, K., Heberle, E., Moosmann, E., & Ruffler, R. (1949). *Unser rechenbuch* Heft 7 [Our thanks, booklet 7]. Stuttgart, Germany: Ernst Klett.

Davis, P. J., & Hersh, R. (1981). *The mathematical experience.* Boston, MA: Birkhäuser.

Dowling, P. (1998). *The sociology of mathematics education: mathematical myths/pedagogic texts.* London, England: RoutledgeFalmer.

Feller, F. E., & Odermann, C. G. (1866). *Das Ganze der kaufmännischen Arithmetik* (10th ed.) [The whole of commercial arithmetic]. Leipzig, Germany: Otto August Schulz.

Gellert, U., & Jablonka, E. (Eds.) (2007). *Mathematisation and demathematisation: Social, philosophical, and educational ramifications.* Rotterdam, the Netherlands: Sense.

Hart, W.W. (1942). *Basic mathematics: A survey course.* Boston, MA: D.C. Heath.

Jablonka, E. (1996). *Meta-Analyse von Zugängen zur mathematischen Modellbildung und Konsequenzen für den Unterricht* [Meta-analysis of approaches to mathematical modelling and consequences for teaching]. Berlin, Germany: transparent.

Lötzbeyer, P., & Molthan, A. (1929). *Rechenbuch für höhere Schulen* , 1. Heft [Thanks for secondary schools, booklet 1]. Dresden, Germany: L. Ehlermann.

Porthe, O. (1937). *A. Büttners Berliner Rechenbuch,* Heft 1 [A. Badr Berlin arithmetic]. Leipzig, Germany: Ferdinand Hirt und Sohn.

Smith, J.H. (1877). *Elementary algebra: Without answers.* London, England: Rivingtons.

Strunz, K. (1953). *Pädagogische Psychologie des mathematischen Denkens* [Educational psychology of mathematical thinking]. Heidelberg, Germany: Quelle & Meyer.

Wartegg, E. (1939). Gestaltung und character [Design and character]. *Zeitschrift für angewandte Psychologie und Charakter-Kunde,* Beiheft 84 [Journal of applied psychology and character customer, Supplement 84].

Uwe Gellert
Freie Universität Berlin
Germany

CHAPTER 6

WHO NEEDS MATHEMATICS EDUCATORS FOR MATHEMATICS EDUCATION, ANYWAY?

Brian Greer

In this chapter, I have three aims. The first is to respond directly to the polemic: Who needs mathematicians for math, anyway (Stotsky, 2009). The second is to comment on the suppression of the field of mathematics education within the deliberations and publications of the National Mathematics Advisory Panel (NMAP) (United States Department of Education, 2008). The third is to weigh the necessary, but not sufficient, contribution of mathematicians to mathematics education.

Echoing Spindler and Spindler (1998, p. 27): "This is a biased paper written from a liberal point of view."

MANHATTAN COCKTAIL

The article by Stotsky (2009) was published, not as an academic publication, but in *City Journal*, a publication of the Manhattan Institute, a conservative

Refractions of Mathematics Education, pages 99–116
Copyright © 2015 by Information Age Publishing
99

think tank (Spring, 2010, pp. 136–141). I assume that Stotsky, by her title, meant: Who needs mathematicians for mathematics education, anyway? Even the revised title is odd, since the contributions of mathematicians to mathematics education are obviously necessary; and it is easy to think of many mathematicians who have so contributed. However, they are certainly not sufficient; and it is equally easy to name names of mathematicians whose influence on mathematics education has been detrimental.

Stotsky had been reacting to the opinions on NMAP expressed by the author of this chapter (Greer, 2008a, 2008b) and others in an issue of the journal *The Montana Mathematics Enthusiast* (TMME). While there may be a degree of ego involved in my response here, I submit that the points I will make have a larger significance. Of necessity, the issues I will address are highly selective. To verify the context of quotations I cite, and see what else Stotsky wrote, one may access the paper on the internet.

Following the Party Line?

Stotsky wrote:

> The mathematics educators' response to the panel's report [NAMP] came as no surprise. The *Montana Mathematics Enthusiast*, a journal put out by an NCTM [National Council for Teachers of Mathematics] state affiliate, was the first to declare the party line in its July 2008 issue, which featured highly critical essays by five mathematics educators.

The characterization of mathematics educators as a homogeneous group is simplistic; the reference to a party line offensive. It is true that I have been a member of NCTM most of the time since I joined in 1985; have spoken at their conferences, and contributed to their publications. Nevertheless, I disagree fundamentally with NCTM on many issues, including their reaction to NMAP. Putting together the papers for TMME was solely my initiative, with the support of its editor, Bharath Sriraman; no contact with NCTM was involved.

Cultural Historical Activity Theory and Constructivism

Stotsky wrote:

> Two theories lie behind the educators' new approach to math teaching: "cultural-historical activity theory" and "constructivism." According to cultural-historical activity theory, schooling as it exists today reinforces an illegitimate social order. Typical of this mindset is Brian Greer, a mathemat-

ics educator at Portland State University, who argues "against the goal of 'algebra for all' on the grounds that... most individuals in our society do not need to have studied algebra." According to Greer, the proper approach to teaching math "now questions whether mathematics as a school subject should continue to be dominated by mathematics as an academic discipline or should reflect more fully the range of mathematical activities in which humans engage." The primary role of math teachers, constructivists say in turn, shouldn't be to explain or otherwise try to "transfer" their mathematical knowledge to students; that would be ineffective. Instead, they must help the students construct their own understanding of mathematics and find their *own* math solutions.

I find it impossible to follow a coherent line of argument in the above paragraph, so I will limit myself to a few details. I conjecture that in the first sentence, Stotsky was not referring to the family of theories going by the term *cultural historical activity theory* (which has limited prominence within the field of mathematics education), but rather, to a more general position that mathematics and mathematics education are historically, culturally, socially, and, indeed, politically embedded activities. With regard to the second theory she cited as central, Stotsky's views on constructivism are on record in another polemic, of which she was one of the authors; it was titled, "Ten myths about math education and why you shouldn't believe them" (nychold.com/myths-050504.html).

I do agree with the statement that schooling, as it exists today, reinforces an illegitimate social order; but elaboration of that position lies outside the scope of this paper.

Algebra for All?

The reader is invited to compare what Stotsky sparingly quoted above with the original passage (Greer, 2008a, p. 428):

> I argue against the goal of "algebra for all" on the grounds that, while collectively a cadre of mathematical and scientific specialists is needed for society to operate effectively, most individuals in our society do not need to have studied algebra. This by no means implies that anyone should be denied the opportunity to do so (in an intellectually stimulating way). Moreover, students should be encouraged to study algebra in the spirit of keeping options open, given its status as a gatekeeper to many educational and economic opportunities.

It is important to note that not all mathematicians disagree with this position. For example, Philip Davis (1999) wrote:

What is necessary is to teach enough so that the commonplace diurnal mathematical demands placed on the population are readily fulfilled. What is also necessary is to infuse sufficient mathematical and historical literacy that people will be able to understand that the mathematizations put in place in society do not come down from the heavens: that they do not operate as pieces of inexplicable ju-ju, that mathematizations are human cultural arrangements and should be subject to the same sort of critical evaluation as all human arrangements. At the risk of sounding like a traitor to my profession, I would say that high school algebra or beyond is not necessary to achieve this goal.

Another mathematician, Lynn Steen, wrote (2004, p. 53):

One of the many ills afflicting mathematics education, is its excessively narrow focus on algebraic symbol manipulation, to the detriment of more widely useful aspects of the mathematical sciences.

Who Needs Understanding for Mathematics, Anyway?

The ongoing arguments in the United States, and elsewhere, about how mathematics should be taught, intensified following the publication of *Curriculum and Evaluation Standards for School Mathematics* (National Council of Teachers of Mathematics, 1989). Exemplary of the grounds on which these arguments had been waged, was the suggestion to give increased attention to aspects such as number sense, meaning of fractions and decimals, use of calculators for complex computation, actual measuring, problem-solving strategies, and justification of thinking (selected from a list of 37 recommendations); and that decreased attention be given to isolated treatment of division facts, paper-and-pencil fraction computation, use of clue words to determine which operation to use (in word problems), and rote memorization of rules (selected from 18 recommendations) (NCTM, 1989, pp. 20–21).

The fault line between positions on such recommendations, has reflected attitudes toward the understanding of mathematics as transcending mere computational and procedural competence. Stotsky stated that: "Math educators proclaimed a brand-new objective—conveniently indefinable and immeasurable—called 'deep conceptual understanding.'" The notion that deep conceptual understanding, as an objective, was "brand-new," betrayed ignorance of the history of the field (Wertheimer, 1945/1982). It is certainly an objective that is difficult to define and to measure, and a great deal of the literature in mathematics education reflects and addresses that difficulty; but are not academics expected to deal with concepts that are difficult to define and to measure? Conversely, computational and procedural competence *are* conveniently definable and measurable; and this convenience is reflected in most standardized mathematics tests, often with pernicious effects.

Understanding may indeed be difficult to define and measure, but it is not difficult to illustrate. One of my favorite examples comes from *Productive Thinking* (Wertheimer, 1945/1982). Wertheimer (p. 130) asked children (the reader is invited to stop and consider this example) to find what number the following expression is equal to:

$$\frac{274 + 274 + 274 + 274 + 274}{5}$$

Anyone who computes a repeated addition, or multiplication, followed by a division, demonstrates computational fluency; but this performance strongly suggests a lack of conceptual understanding. Wertheimer related his surprise that, while most of the bright subjects he asked "enjoyed the joke" (p. 112), "a number of children who were especially good at arithmetic...were entirely blind" (p. 113).

As a second example, I posed this question to future elementary teachers studying the concept of slope:

A candle, initially 24 cm high, is burning *down at the rate of 3 cm per minute.* If you plot the graph of height of the candle against time, what will be the slope of the line?

Most of the class, competently but unnecessarily, drew the graph and computed the slope (rise over run). But if you *understand* how a graph of this type represents constant rate of change, you will realize that the answer (−3) is directly given in the question, in the phrase italicized.

By way of a third example, this is a *Scholastic Aptitude Test* problem recently posted on a blog (rationalmathed.blogspot.com/). Again, the reader is invited to find the answer before proceeding:

The stopping distance of a car is the number of feet that the car travels after the driver starts applying the brakes. The stopping distance of a certain car is directly proportional to the square of the speed of the car, in miles per hour, at the time the brakes are first applied. If the car's stopping distance for an initial speed of 20 miles per hour is 17 feet, what is its stopping distance for an initial speed of 40 miles per hour?

(A) 34 feet   (B) 51 feet   (C) 60 feet   (D) 68 feet   (E) 85 feet

The first pointer that the author of this question lacks conceptual understanding of arguments based on dimensionality and proportionality, is the unnecessary "in miles per hour"; the statement would be unaffected by changing that to "in millimeters per century." The main point, however, comes in the official explanation of how to find the right answer, which involves defining a number of variables, setting up a number of equations, and carrying out a number of calculations involving the number 0.0425 (you might like to figure out where that comes from). By contrast, if you understood proportionality and dimensionality in this context, and could spot the conveniently simple ratio between 20 and 40, then it would be immediately clear that the answer is 17×2^2.

Research and Improving Mathematics Education

Stotsky wrote:

> Issue editor Greer declared in his overview that the panel's report offered nothing useful, since it had "restricted" itself to scientific research and ignored the "rich reflections" of educators, who, in his judgment, had produced the "deepest work in the field."

Nowhere did I opine that the panel's report offered nothing useful, or make any equivalent statement. Indeed, I believe that it offered much that was useful, above all in giving insight into the political and ideological nature of mathematics education (Greer, 2012b). The section in my paper to which Stotsky was referring was the following:

> In particular, the members (with commendable, but arguably misplaced, diligence) plowed through (and regurgitated) huge masses of empirical work, preselected according to strict criteria that excluded most of the deepest work in the field. In my opinion, a lot of their time and intellectual energy would have been much better spent on reading rich reflections on mathematics education, such as the work of Hans Freudenthal (1983, 1991).

In my opinion, NMAP was fatally flawed from its inception by operating (except when it was expedient to appeal to the members' opinions) on the principle that if one cannot do randomized experiments, then one must be silent. Some research is useful to mathematics education, and much is not. For it to be useful requires interpretation, synthesis, reflection, and contextualization; all of which respects I find NMAP lacking.

Politics of (Mathematics) Education in the United States

There is obviously much more I could write on differences of opinion about how mathematics should be taught in schools. A full treatment would require an analysis of the state of educational politics in the United States (Spring, 2010), including the existential threat to public schooling (Ravitch, 2010, 2013); and the pervasive rhetoric that casts mathematics and science education as a matter of national security and continuing military and economic world dominance by the United States (Klein & Rice, 2012).

DISAPPEARING TRICK

In April 2006, President Bush established NMAP, which issued its final report in March 2008. There is admirably complete public-domain documentation on the workings of the panel,[1] which I hope is providing the material for historians and sociologists of education to analyze what is a fascinating case study in the politics of education (Greer, 2012b).

The first item in the list of recommendations for the panel, referred to "the critical skills and skill progressions for students to acquire competence in algebra and readiness for higher levels of mathematics" (p. 71). Kilpatrick (2009, p. 393) wondered:

> Did George W. Bush, having decided that he needed to convene a panel to advise him on how the United States might implement the policy of fostering "greater knowledge of and improved performance in mathematics among American students" (p. 71) wake up one morning and say, "Competence in algebra ought to be the first order of business"? As Danny Martin (2008) notes, the president's choice of algebra was far from politically neutral, and one can raise questions about it: Why algebra? Who decides? Algebra for whom and for what not-so-apparent purposes? Whose interests are served by these choices? Whose interests are not served?

In the absence of inside information, my conjecture is that certain politically active mathematicians, who had a particular view of mathematics and how it should be taught, had the ear of the president. In any case, the priority given to algebra was maintained throughout the exercise.

The organization, the reports, and the recommendations from the panel have been heavily criticized by mathematics educators on many counts in *The Montana Mathematics Enthusiast*, Vol. 5 (2&3), and in *Educational Researcher*, Vol. 37(9). Here I will concentrate on the ways in which mathematics education as a field has been rendered all but invisible in NMAP.

Mechanisms of Exclusion

First, consider the make-up of the panel: Confrey, Maloney, and Nguyen (2008, p. 631) characterized it as "unusual, considering the charge. " They pointed out that only five out of 19 members had sustained interactions with mathematics instruction at the K–12 level, and fewer than half the panel members (not including Stotsky) had documented academic preparation in mathematics. (Who needs mathematics for advising on policy for mathematics education, anyway?) As Confrey et al. also point out, the panel lacked representatives of the humanistic disciplines that have enriched the field of mathematics education in recent decades.

Yet more extreme omissions characterized the literature cited by the panel. I searched the report for references to the work of the leading scholars in our field; almost without exception, there were none. In particular, given the dominance of algebra, the absence of reference to the work of Jim Kaput (Kaput, 1999) and others was a scandal. Confrey et al. (2008, p. 633) reported that only slightly more than 10% of the 466 journal articles referenced in the *Report of the Task Group on Learning Processes* (which included *no* mathematics educators) (Geary et al., 2008) were from mathematics education journals; whereas articles published in journals on psychology, child development, and educational psychology constituted about 70% of the references.

The most extreme manifestation of exclusion has been with the second *Handbook of Research on Learning and Teaching Mathematics* (Lester, 2007), an NCTM publication. While the panel was ongoing, Frank Lester offered to make the chapters available; his offer was ignored, save for a formal acknowledgment of receipt. The denial of the significance of the handbook, as an important summary of research in the field, was perverse.

Where's the Math?

A second disappearing trick concerns mathematics itself. We could ask: Who needs mathematics, apart from arithmetic, algebra, and calculus, for school mathematics, anyway? This would draw attention to the remarkable mathematical omissions in the NMAP report of geometry, probability, data handling, mathematical modeling, and applications, which are all barely mentioned. Moreover, a visitor from Mars who reads the report, could be forgiven for concluding that earthlings have not found any reason for using computers in mathematics education beyond drill and practice.

Should mathematicians be happy with NMAP? I suspect that many mathematicians, if asked what is the most important element of doing mathematics, would say either proof or problem-solving. But NMAP is largely silent on both. I am not aware of prominent mathematicians (certainly not those on the panel) who have criticized it on these grounds; but my antennae on this aspect are weak, so I would welcome information.

I have not even touched on the failure of NMAP to mention vital contextual embeddings, such as the history of mathematics (social and cultural as well as intellectual); a discussion of the roles of mathematics in society; or the inadequate attention to equity and diversity issues (Martin, 2008).

Response to NMAP from Mathematics Educators

After the report was published, mathematics educators responded strongly in the issues of *The Montana Mathematics Enthusiast* and *Educational Researcher,* which was referred to earlier. However, I have been disappointed that mathematics educators who were on the panel (including Skip Fennell, then president of NCTM), failed to register any kind of significant public protest of which I am aware (I would be very happy to be informed otherwise). For whatever reasons, they declined to take any of a number of possible actions; such as submiting a minority report, or even resigning in protest. I think they have let the field down.

The response by NCTM I found disappointing. (But then I am unaware of the *realpolitik* in which they had to operate, and intense pressure they have been under since 1989). In particular, I am unaware of any protest by NCTM about the rejection of the handbook (Lester, 2007). (Again, I would be happy to be informed otherwise). It is interesting that the current president of NCTM made a statement (Shaughnessy, 2010) in which he argued for "statistics for all." It is also noteworthy, despite his leadership for many years in research on teaching and learning probability and statistics, that a search for "Shaughnessy" in the NMAP reports yields not a single hit. And, there is the irony (which has been appreciated by Shaughnessy in a conversation I had with him) in the fact that NMAP essentially ignored statistics as a branch of mathematics; this leads one to ask where will all the statisticians come from to provide the expertise for all that methodologically rigorous research?

So What?

I took NMAP personally as an attack on the field in which I have spent my career. I have been told by respected colleagues, that the NMAP report will gather dust on a shelf; move on, get over it, has been their advice. I am not so sanguine that it will not continue to have an influence; and in any case, the whole enterprise merits analysis from the point of view of sociology and the politics of education (Greer, 2012b). True, it may be of only academic importance against the background of what is happening now in educational politics in the United States and globally. Flaubert, in 1872, writing to Turgenev about the proposed reform of public education in France, declared that "I have always tried to live in an ivory tower, but a tide of shit is beating at its walls, threatening to undermine it" (www.rjgeib. com/about-me/faq/gustave-flaubert.html).

MATHEMATICIANS FOR MATHEMATICS EDUCATION

Throughout the development of mathematics, and especially in modern times, there have been many mathematicians who have offered suggestions for mathematics education. Recent outstanding examples of mathematicians in the United States who have contributed greatly to mathematics education include George Polya, Bob Davis, and Jim Kaput (all ignored in NMAP). There have been major historical episodes in which mathematicians have significantly impacted mathematics education, notably the new mathematics of the 1960s (which took different forms in different educational regimes), and the ongoing conflict of views generally referred to as the "math wars" (for reasons why I think this metaphor is pernicious, see Greer, 2012b, p. 109). In both of these cases, from my liberal point of view, I consider that the overall impact of mathematicians has been negative.

In this section, I will focus on a pre-eminent example of a mathematician contributing positively to mathematics education. Then I will comment on the assumption too often made by mathematicians (and others), that expertise in a discipline affords authority for teaching that discipline. I will critique the assumption of a straightforward relationship between mathematics-as-discipline and mathematics-as-school-subject, and finish with the most fundamental, but rarely posed, question: Who needs mathematics, and for what, anyway?

Hans Freudenthal (1905–1990)

Freudenthal was a very considerable mathematician (particularly in the fields of topology, geometry, and Lie groups), before he turned his attention to mathematics education. He considered one of his greatest achievements to have ben quarantining the Netherlands from the effects of the new math movement; for which he wrote a suitable epitaph: "New math's wrong perspective was to replace the learner's insight with the adult mathematician's" (Freudenthal, 1991, p. 112). He regarded the notion of an architecture of mathematics, which could be immaculately transmitted to students through a logical progression, as a didactical inversion; it ignored the fundamental reality of mathematics as passing through a long process of development, and accordingly, requiring re-invention under instruction as pedagogy. Emphatically, this does not mean that the learner should re-invent anything, which is patent nonsense. Rather, in Freudenthal's words (1991, p. 49): "the learner should reinvent mathematising rather than mathematics; abstracting rather than abstractions; schematizing rather that schemes; formalizing rather than formulae; algorithmising rather than algorithms, verbalizing rather than language . . ."

Freudenthal also offered a profound critique of Piaget's excessive concern with logic; by contrast, he looked for the origins of mathematics in human experience. Piaget famously believed he could discern an alignment between the structures posited in his theory of cognitive development, and those developed by the Bourbaki group, as a formal architecture for mathematics. On this conceit, Freudenthal (1973, p. 46) commented, with typical forcefulness, as follows:

> The most spectacular example of organizing mathematics is, of course, Bourbaki. How convincing this organization of mathematics is! So convincing that Piaget could rediscover Bourbaki's system in developmental psychology. Poor Piaget! He did not fare much better than Kant, who had barely consecrated Euclidean space as a "pure intuition" when non-Euclidean geometry was discovered! Piaget is not a mathematician, so he could not know how unreliable mathematical system builders are.... Mathematics is never finished—anyone who worships a certain system of mathematics should take heed of this advice.

He also offered trenchant critiques of the shallowness and simplism of much of experimental psychology and psychometrics (Freudenthal, 1978).

In 1971, Freudenthal established the Institute for Development of Mathematics Education in Utrecht, now renamed the Freudenthal Institute. With Adriaan Treffers and many other colleagues, he worked to create an integrated approach to development and research, including all aspects of teaching mathematics that had, and continues to have (Van den Heuvel-Panhuizen, 2010), a major impact in the Netherlands, and indeed worldwide.

Finally, I cannot resist quoting what he said about bodies such as NMAP, which are set up to advise policymakers: "They are not used for anything else than rationalizing a politically based decision" (Freudenthal, 1991, p. 150).

Caste System

There is a tendency among mathematicians, by no means alone among discipline specialists, to assume that expertise in the discipline automatically affords expertise in how to teach it. Consider the following quotation (Greer & Mukhopadhyay, in press), amended—in a way that should be clear—from Macedo, Dendrinos and Gounari (2003, pp. 19–20):

> The asymmetrical power relations between ~~literature~~ *mathematics* and ~~language~~ *mathematics education* studies reproduce the false notion that anyone trained in ~~literature~~ *mathematics* is automatically endowed (through osmosis) with the necessary skills to teach ~~the language in which the literature is written~~ *mathematics*. This position precludes viewing ~~language~~ *mathematics teaching* as a complex field of study which demands rigorous understanding of

theories of ~~language acquisition~~ *learning mathematics* coupled with a thorough knowledge of the ~~language~~ *mathematics* being taught and its functions in the society that generates and sustains it.

When Lynn Steen, a mathematician of considerable standing, was asked in an interview[2] for explanations of the conflict within mathematics education, he suggested three:

> One is the natural tendency of parents to want their children to go through the same education that they received—even when, as often is the case with mathematics, they admit that it was a painful and unsuccessful ordeal. [...] Another source was scientists and mathematicians who pretty much breezed through school mathematics and who were increasingly frustrated with graduates (often their own children) who did not seem to know what these scientists knew (or thought they knew) when they had graduated from high school. [...]

> A third source can be traced to the way in which the NCTM *Standards* upset the caste system in mathematics education. Mathematicians are accustomed to a hierarchy of status and influence with internationally recognized researchers at the top, ordinary college teachers in the middle, below them high school teachers, and at the very bottom teachers in elementary grades. The gradient is determined by level of mathematics education and research. So it came as somewhat of a shock to research mathematicians when the organization representing elementary and secondary school teachers, seemingly without notice or permission, deigned to issue "standards" for mathematics.

Forms of Mathematical Practice

Mathematics as a contemporary discipline is predominantly taken as essentially well defined and universal. This perception can be contested, however. For example, Raju (2007) has argued that there are two streams of mathematics: "(1) from Greece and Egypt a mathematics that was spiritual, anti-empirical, proof-oriented, and explicitly religious; and (2) from India via Arabs, a mathematics that was pro-empirical, and calculation-oriented, with practical objectives" (p. 413). And, as Raju argued, the emergence of the computer and related tools shifted that balance.

The claim of the universality of mathematics is typically followed by an appeal to the self-evidence of, for example, the angles of a triangle adding up to half a complete rotation, or $2 + 2 = 4$, wherever you are and across eternity. However, it is by no means self-evident if one considers contested areas of mathematics, such as probability theory or non-standard analysis. Raju (2007) clarified that the claimed universality of Western mathematics depends on the choice of a particular two-valued logic, which is a cultural decision and does not provide a viable model for all applications of mathematics.

Mathematicians and others tend to assume a straightforward relationship between mathematics-as-a-discipline and mathematics-as-school-subject; whereby the purpose of the latter is to transmit the basic knowledge of the former to all students, from which the next generation of mathematicians will emerge. However, shaping mathematics education for all in relation to the needs of mathematicians to reproduce their kind, has negative side-effects; in particular the alienation felt by many toward school mathematics that they (generally correctly) see as remote from their lived experience. From Eva Jablonka I first heard the analogy that striving to have all students learn a considerable amount of formal mathematics in order to maximize the chance of a strong new generation of mathematicians, is rather like making every child play a certain sport with the aim of producing a strong national team in that sport.

I argue (Greer, 2012a) that there are other forms of mathematical practice that should contribute to the framing of school mathematics. Since Ubiratan D'Ambrosio (1985) introduced the concept of ethnomathematics, there has been considerable work on non-academic mathematical practices that occur in essentially every culture; also in occupational groups such as carpenters. Another aspect of ethnomathematics is deconstructing the master narrative, which claims that real mathematics began with the Greeks, lay dormant for a thousand years, and reappeared in the Italian Renaissance (Joseph, 1992; Raju, 2007). How the ethnomathematical perspective might be accommodated within school mathematics, is as yet under theorized; it remains an open question (Pinxten & Francois, 2011; Greer, 2013).

Another family of mathematical practices, the importance of which for school mathematics needs to be considered, is termed "mathematics in action" by Ole Skovsmose (Skovsmose, 2005, pp. 79–91). He means the applications of mathematical models in ways that format our lives, often without our being aware of their existence and almost always beyond our control. In order for future citizens to have a disposition to want to understand this situation, and the agency to be other that powerless in the face of it, students should learn about the nature, and particularly the limitations, of mathematical modeling. They should start early with the key insight that physical and social phenomena are not unproblematically modeled by arithmetic operations.

Pinxten and Francois (2011, p. 271) have proposed a balanced approach, based on the characterization of ethnomathematics as embodying all forms of mathematical practice, including academic mathematics:

On the one hand, we stress that the lower limit mathematical literacy is or should be a condition to guarantee access to more or even maximal mathematical expertise for all. On the other hand ... we invite scholars to support the strong claim that a variety of mathematical insights and relevant knowledges are feasible, and that it is a debatable *political* choice to favor one over the next one. (Emphasis added)

What is Mathematics Education *For*?

Consider the way in which the National Council of Teachers of Mathematics (NCTM) began its draft of Standards 2000. No Socrates-like character asked: And shall we teach mathematics? Even if the answer was a preordained, (of course, Socrates) asking the question raised a host of others: To whom shall we teach mathematics; for what ends; mathematics of what sort; in what relation to students' expressed needs; in what relation to our primary aims; and what are these aims? (Noddings, 2003, p. 87)

Traditional reasons for teaching mathematics include: "The need to produce another generation of scholars to continue advancing the discipline of mathematics; the supply of a cadre of scientists and others such as engineers who need strong mathematical competence; as a training in thinking, problem solving, and creativity ... ; as part of cultural heritage as much as literature or music" (Greer, 2009, p. 3). In a trenchant critique, Pais (2009, p. 57) argued that education in general, and mathematics education in particular, "has developed the function of reducing, dominating, and suffocating education by the way of re-inscribing it within the structure of the state." By contrast, there is the possibility of mathematics education as liberatory, the best example being the work of Gutstein (2012) in Chicago, in the spirit of Paulo Freire (Freire and Macedo, 1987).

In recent years—most obviously, *but by no means exclusively*, in the United States—the purpose of mathematics (and science) education has been framed in nationalistic terms, as relating to economic and even military competitiveness. In the executive summary of the final report (NMAP, 2008, p. xi), it is stated that:

During most of the 20th century, the United States possessed peerless mathematical prowess ... But without substantial and sustained changes to its educational system, the United States will relinquish its leadership in the 21st century.

Much of the commentary on mathematics and science in the United States has focused on national economic competitiveness and the economic well-being of citizens and enterprises. There is reason enough for concern about these matters, but it is yet more fundamental to recognize that *the safety of the nation and the quality of life*—not just the prosperity of the nation—are at issue. (Emphasis added)

In the same vein, is a recent report emanating from The Council for Foreign Relations titled *The U.S. Education Reform and National Security*. The chief authors were Condoleeza Rice (secretary of state under President George Bush) and Joel Klein, a lawyer who was put in charge of education in New York City for eight years (for a critique of his regime, see Ravitch, 2010, pp. 69–91). Klein now works for Rupert Murdoch's *News International* in charge of their education division, and enjoys close ties with the British government.

The subordination of mathematics education to nationalism that is illustrated by these examples, stands in stark contrast to the plea by Ubi D'Ambrosio (2010), which was that mathematicians and mathematics educators have ethical responsibilities in relation to the most important problem facing humankind; namely survival with dignity.

TO SUMMARIZE

- Academic mathematics is multicultural, and there are pancultural mathematical practices beyond the academic.
- Mathematics education, like all education, is political. In view of the pervasive influence of applications of mathematics in the contemporary world, mathematics educations have particular ethical responsibilities.
- Improving mathematics education is a human problem, not a technical problem (Kilpatrick, 1981).
- Mathematicians have a vital role to play, but mathematics education is much too important to leave primarily to mathematicians.

NOTES

1. See http://www2.ed.gov/about/bdscomm/list/mathpanel/index.html
2. mathnotations.blogspot.com/2007/09/interview-with-prof-lynn-arthur-steen_14.html

REFERENCES

Confrey, J., Maloney, A. P., & Nguyen, K. H. (2008). Breaching the conditions for success for a National Advisory Panel. *Educational Researcher, 37*(9), 631–637.

D'Ambrosio, U. (1985). Ethnomathematics and its place in the history and pedagogy of mathematics. *For the Learning of Mathematics, 5*(1), 44–48.

D'Ambrosio, U. (2010). Mathematics education and survival with dignity. In H. Alro, O. Ravn & P. Valero (Eds.), *Critical mathematics education: Past, present, and future* (pp. 51–63). Rotterdam, the Netherlands: Sense.

Davis, P. (1999, March 22). Testing: One, two, three; Testing: One,... can you hear me back there? *SIAM News.* Retrieved from http://www.siam.org/news/news. php?id=719

Freire, P., & Macedo, D. (1987). *Literacy: Reading the word and the world.* South Hadley, MA: Bergen & Garvey.

Freudenthal, H. (1973). *Mathematics as an educational task.* Dordrecht, the Netherlands: Reidel.

Freudenthal, H. (1978). *Weeding and sowing: Preface to a science of mathematics education.* Dordrecht, the Netherlands: Reidel.

Freudenthal, H. (1983). *Didactical phenomenology of mathematical structures.* Dordrecht, the Netherlands: Reidel.

Freudenthal, H. (1991). *Revisiting mathematics education.* Dordrecht, the Netherlands: Kluwer.

Geary, D. C., Boykin, A. W., Embretson, S., Reyna, V., Siegler, R., Berch, D. B., et al. (2008). "Report of the task group on learning processes." Retrieved from http://www.ed.gov/about/bdscomm/list/mathpanel/report/learning-processes.pdf

Greer, B. (2008a). Algebra for all? *Montana Mathematics Enthusiast, 5*(2&3), 423–428.

Greer, B. (2008b). Reaction to the final report of the national mathematics advisory panel. *Montana Mathematics Enthusiast, 5*(2&3), 365–370.

Greer, B. (2009). What is mathematics education for? In P. Ernest, B. Greer, & B. Sriraman (Eds.), *Critical issues in mathematics education* (pp. 3–6). Charlotte, NC: Information Age.

Greer, B. (2012a, June). *Mathematics education and mathematics: Relationships among activity systems.* [Paper presented at Royal Anthropological Institute Conference on Anthropology in the World]. London, England.

Greer, B. (2012b) The United States National Mathematics Advisory Panel as political theatre. In O. Skovsmose & B. Greer (Eds.), *Opening the cage: Critique and politics of mathematics education* (pp. 107–124). Rotterdam, the Netherlands: Sense.

Greer, B. (2013). Teaching through ethnomathematics: Possibilities and dilemmas. In M. Berger, K. Brodie, V. Frith, & K. le Roux (Eds.), *Proceedings of the 7th International Mathematics Education and Society Conference* (pp. 94–102). Cape Town, South Africa: MES7.

Greer, B., & Mukhopadhyay, S. (in press). The hegemony of English mathematics. In P. Ernest & B. Sriraman (Eds.), *Critical mathematics education: Theory, praxis and reality.* Charlotte, NC: Information Age.

Gutstein, E. (2012). Mathematics as a weapon in the struggle. In O. Skovsmose & B. Greer (Eds.), *Opening the cage: Critique and politics of mathematics education* (pp. 23–48). Rotterdam, the Netherlands: Sense.

Joseph, G. G. (1992). *The crest of the peacock: Non-European roots of mathematics.* London, England: Penguin.

Kaput, J. J. (1999). Teaching and learning a new algebra. In E. Fennema & T. A. Romberg (Eds.), *Mathematics classrooms that promote understanding* (pp. 133–155). Mahwah, NJ: Lawrence Erlbaum.

Kilpatrick, J. (1981). The reasonable ineffectiveness of research in mathematics education. *For the Learning of Mathematics, 2*(2), 22–29.

Kilpatrick, J. (2009). Algebra is symbolic. In S. L. Swars, D. W. Stinson & S. Lemons-Smith (Eds.), *Proceedings of the 31st annual meeting of the North American Chapter of the International Group for the Psychology of Mathematics Education* (pp. 11–21). Atlanta, GA: Georgia State University.

Klein, J., & Rice, C. (2012). *U.S. education reform and national security.* Center for Foreign Relations: Independent Task Force Report 68. Retrieved from www.cfr.org/united-states/us-education-reform-national-security/p27618

Lester, F. K., Jr. (Ed.) (2007). *Second handbook of research on mathematics teaching and learning.* Charlotte, NC: Information Age.

Macedo, D., Dendrinos, B., & & Gounari, P. (2003). *The hegemony of English.* Boulder, CO: Paradigm.

Martin, D. B. (2008). E(race)ing race from a national conversation on mathematics teaching and learning: The National Mathematics Advisory Panel as white institutional space. *Montana Mathematics Enthusiast, 5*(2&3), 387–398.

National Council of Teachers of Mathematics (1989). *Curriculum and evaluation standards for school mathematics.* Reston, VA: Author.

National Mathematics Advisory Panel (2008). *Foundations for success. The final report of the National Mathematics Advisory Panel.* Washington, DC: U.S. Department of Education.

Noddings, N. (2003). *Happiness and education.* New York, NY: Cambridge University Press.

Pais, A. (2009). The tension between what mathematics education should be for and what it is actually for. In P. Ernest, B. Greer, & B. Sriraman (Eds.), *Critical issues in mathematics education* (pp. 53–60). Charlotte, NC: Information Age.

Pinxten, R., & Francois, K. (2011). Politics in an Indian canyon? Some thoughts on the implications of ethnomathematics. *Educational Studies in Mathematics, 78,* 261–273.

Raju, C. K. (2007). *Cultural foundations of mathematics: The nature of mathematical proof and the transmission of the calculus from India to Europe in the 16th c. CE.* New Delhi, India: Centre for Studies in Civilizations/ PearsonLongman.

Ravitch, D. (2010). *The death and life of the great American school system: How testing and choice are undermining education.* New York, NY: Basic Books.

Ravitch, D. (2013). *Reign of error.* New York, NY: Knopf.

Shaughnessy, J. M. (2010, August). "Statistics for all—the flip side of quantitative reasoning." *Summing Up.* Retrieved from http://www.nctm.org/about/content.aspx?id=26327

Skovsmose, O. (2005). *Traveling through education: Uncertainty, mathematics, responsibility.* Rotterdam, the Netherlands: Sense.

Spindler, G., & Spindler, L. (1998). Cultural politics of the white ethniclass in the mid 90s. In Y. Zou & E. T. Trueba (Eds.), *Ethnic identity and power: Cultural contexts of political action in schools and society* (pp. 27–47). Albany, NY: SUNY Press.

Spring, J. (2010). *Political agendas for education.* New York, NY: Routledge.

Steen, L. A. (2004). Data, shapes, symbols: Achieving balance in school mathematics. In B. Madison & L. A. Steen (Eds.), *Quantitative literacy: Why numeracy matters for schools and colleges* (pp. 53–74). Washington, DC: Mathematical Association of America.

Stotsky, S. (2009, November 13). Who needs mathematicians for math, anyway? *City Journal.* Retrieved from www.city-journal.org/2009/eon1113ss.html

United States Department of Education (2008). *Foundations for success: The final report of the National Mathematics Advisory Panel.* Washington, DC: Author.

Van den Heuvel-Panhuizen (2010). Reform under attack— forty years of working on better mathematics education thrown on the scrapheap? No way! In L. Sparrow, B. Kissane & C. Hurst (Eds.), *Proceedings of 33rd Annual Conference of the Mathematics Education Research Group of Australasia.* Fremantle, Australia: Author. Retrieved from www.fisme. science. uu.nl/staff/marjah/download/ VdHeuvel-Panhuizen_keynote_ MERGA33co.pdf

Wertheimer, M. (1945/1982). *Productive thinking.* Chicago, IL: University of Chicago Press.

Brian Greer
Portland State University
USA

CHAPTER 7

PURSUING REFRACTIONS

A Conversation with Eva Jablonka

Steve Lerman

The following conversation took place at King's College London, England, on April 29, 2014.

Steve: The title of this book, which I understand you know not a lot about, I think is a really nice title: *Refractions of mathematics education.* I thought I'd just ask you what sort of images or ideas does the title call up for you?

Eva: It can be taken as a position toward the field of research. I think there is an important activity for researchers, which is, if you assume that you are searching for some sort of "truth," that you need to always ask new questions as soon as some things become an official truth. Then I think there is a need for refraction.

Steve: So refraction is a turn?

Eva: Yes. Refraction could also be interpreted in relation to practice, because research is a completely different activity than teaching. What researchers can do: they can look at a common practice, but this is not to mirror or reflect it; it is a refraction. It is opening up new ways of noticing things and deconstructing common

Refractions of Mathematics Education, pages 117–129

sense interpretations, common sense truths. I can also see this
in the refraction metaphor.

Steve: For me the idea of which medium is it that things are being
refracted through, was meaningful.

Eva: Yes, what they are being refracted through is theories. So the
very act of theorizing is an act of refracting.

Steve: So the book is an invitation to whoever to look at something
new, is that what you say, new kind of theories, or through theories, new ways of thinking about things.

Eva: Hm, but I don't think that there is a privileging of *a priori*
theory in my view, because theorizing is an interaction with the
empirical. But it is also an interaction with other ways of seeing
things. I mean, other theoretical frameworks or attempts of
theorizing. But the framework is not there fully developed right
form the start. It is more a dynamic process, how I see it.

Steve: Right, and you started up by saying that's the work of
researchers?

Eva: That's the work of researchers, but it is then also the responsibility of researchers to give the teachers, whatever practice you
are dealing with, the tools to do their own refractions instead of
didacticizing research results, as if they were unwise students.

Steve: OK, so in terms of the relationship between theory and practice then, what you are suggesting is that it's not a business of
researchers didacticizing what they are doing, but working with
teachers to make sense of . . . those kind of ideas.

Eva: Yes, there's always two possibilities. The one obviously very commonly chosen is to transform their outcomes, theories, frameworks into something more easy. This is the one I would not
choose, because my goal is to create reflection and refractions
of practices.

Steve: You and I have interacted over a number of things, in particular I
was thinking of the project which you had with Christine [Knip-
ping], David [Reid] and Uwe [Gellert], about the achievement
gap, and bringing some theory to analyzing it. So, where are you
at in that project, and in particular, what we just been talking
about, how does that fit with where you are in your work?

Eva: This work, I think, has proceeded quite well. I've always liked
to look into classrooms, because there eventually the meanings
are created and there social structures are reproduced. So the
question was simply about how would they be reproduced or
produced. There is certainly a lot of knowledge about social
and cultural reproduction from the point of view of different
theoretical framings. But the issue here was to put out some

principles, and I think we found different mechanisms in the different contexts we looked at. We have chosen an egalitarian or more inclusive context and a more selective or elitist one in different systems, and classes with different student achievements, which were more or less homogeneous. We looked into the first three weeks when a new group comes together and asked, "Is there anything we can see where the process starts, where some student sticks out, someone is praised?" So it is an analysis of key incidents. We tried to frame this in relation to the different pedagogies in operation, and we roughly started seeing them in terms of framing and classification. We needed the notion of framing, because we wanted to differentiate between sequencing and pacing criteria, otherwise classification and framing coincide. In each of these different contexts, we found a mechanism for hierarchizing students within that particular pedagogic practice. In one highly individualized weakly framed classroom practice in Sweden, the mechanism was about students' choices, one might say about students' cultural capital. Who gets the most out of the teacher and asks all these intelligent questions? If they are just sitting there and calculate something with the textbook, and the teacher is a resource you can ask or not ask, then there is no pedagogic transmission, I think. Framing in terms of Bernstein was marginalized; this is all about the students. There was also a strongly framed and classified, but completely proceduralized, practice in a low achieving class in Germany. Hauke [Straehler-Pohl] has written nicely about that. The content is trivialized, and so there is no distribution of different forms of knowledge, but only the students' discipline. There is only a regulative discourse creating the hierarchy. In the Swedish classrooms we found that a teacher differentially employed generalizing and localizing strategies. Students who asked the teacher for help with the same tasks, got different answers, which were not based on any obvious difference in the ways their questions were posed. But some got answers like: Look, it's like you buy one coke and then you must imagine you buy two cokes; how much would you pay? And others [were] like: You see, you can call this x and this other y. We found that localizing strategies were more likely to be employed for girls. In each of the settings, we see different mechanisms that are optimized towards that particular pedagogic practice, to create a hierarchy among students.

Steve: That's really interesting. I mean the different countries and the different contexts and the different things that you found;

and developing an appropriate rich sociological description. Just coming back to what you were saying before, though, what opportunity has there been to engage the teachers in looking at their practice and...

Eva: When I was in the north of Sweden still, where the density of teachers per square kilometre is quite low, we organized an event called, Hur blir man bäst i klassen?—How does one become the best student in the classroom? And there were 70 teachers registering, a large number. We showed transcripts from a Swedish classroom, and we explained how we see a localizing-generalizing distributive strategy, or excerpts from other classrooms, and our theorizing of it. The teachers were listening and engaged in an open discussion, went into lunch break, came back and it went on and on. Some teachers said: Yes, I think I'm doing this, oh God I'm doing this; But I'm doing it with the girls I think; or, it's awful. The teachers also engaged in a theoretical discussion, and there was no difference in this to a discussion with colleagues. Colleagues certainly would have been more likely to criticize the approach though. I have always shared my research with teachers, and it works particularly well with examples of classroom tasks and transcripts. You don't need any translation, and you don't need any answer. It's just that they learn to notice something, which they haven't been able to notice before, because they did not have the framework or the theory to notice it.

Steve: I'd love to see what happens in their classrooms when they get back afterwards, whether you had a chance to do so. I've also been in situations [like] the ones that you have. We're talking to teachers about some of the ideas and sharing examples. Really, they are stunned and really engaged in it, and [they] want to do something with it. What about other projects that you have been involved in?

Eva: In terms of bigger projects one is this learner's perspective study: a study that started as a critique of the TIMSS video study with three countries, and then has grown into 15 participating countries. [There] were eighth-grade classrooms, and in this I was very much involved, because I spent one year in Australia at the International Centre for Classroom Research, which wasn't there before I came, but I helped to set it up.

Steve: That was in Melbourne with David Clarke. So you did lots on that, you were coordinator of the German group.

Eva: I did my habilitation in the form of a thick report on my analysis of data from the learner's perspective study. I looked at Hong

Kong, [the] United States and Germany; and this [had] a substantial amount of classroom video data and transcripts, student
interviews and students' work, and teacher interviews. There is
a tremendously rich data set in this study. Even if the data is getting old, there's a huge potential of still using this data for testing some theoretical framings, or if you want to develop codes.
In the study, I was interested in the similarities across these
diverse settings: different curriculum traditions, different pedagogic strategies, different achievement levels of students. There
is no specific comparative research, as all research is about
comparing; neither is there much sense in seeing a classroom
practice as significant of a country or any geographical bubble
on a map. It is more interesting to identify the similarities,
because who would not expect differences; and if you zoom in,
everything becomes different. I was always trying to identify and
explain similarities, and that's why I looked across these diverse
contexts. To explain similarities, one needs structural theories; if
one wants to explain differences, one can draw on ethnography.
But one should avoid [producing] explanations that essentialize
culture. This is how I see this work for myself.

Steve: Right, so what kind of things did you find?

Eva: I produced some thick descriptions of the pedagogies, and
everything that the students said about lesson structure to start
with. Within this diversity, one interesting thing was the limited
amount and quality of mathematical reasoning across these
settings. I used a description of mathematical reasoning, which
was open [and] not an *a priori* description of what mathematics
is, but looked when someone gave some reason for doing or saying anything. And then looked [at] what sorts of reasons these
were; for example backing up some algorithm [by] drawing
on a general rule, or justifying the definition of a geometrical
figure. Generally these were not common, especially if I looked
at students' contributions that were not reasoning on request
by the teacher (explain your thinking or tell me why you've
done this), but reasonings that were on their own initiative.
Prophylactic reasoning, one would think, is part of the activity of mathematicians' mathematics; where one doesn't only
give reasons in the case of a dispute, but usually gives reasons
because one wants to argue for the constructions one is producing. In these classrooms it was the other way round; reasons
were mostly given when something went wrong. The teachers'
questions about why a student did this or that signified: Oh,
there must be something wrong. It resembled an everyday

routine of interaction, where you are only to give a reason if you break some rule. If you had been half an hour late, I think, you would have produced a reason why you were delayed. But when everything runs smoothly, we don't back up whatever we are saying or doing all the time. The same asymmetry of reasoning operated in the classrooms, and there's of course the unavoidable asymmetry between teacher and students. This means there are routines in operation that do not suggest student self-initiated mathematical reasoning. As I said, there was not much difference across these contexts except for some exceptional lessons. The lowest amount was in Hong Kong, and there one could look at the authority relationships. So there is a point in looking how cultural traditions merge with pedagogic strategies. Across the countries I looked at, there was one German classroom that was very similar to one of the Hong Kong classrooms. So the similarities came not in packages in terms of countries. The modes of reasoning were dependent on how hierarchical rules, and turn taking mechanisms in these classrooms, operated and how lessons were structured. There's a huge potential in using cross-cultural data in this way. I wrote basically most [about] this in the habilitation report.

Steve: These are important issues, I mean this is really nice stuff, interesting stuff, and not what one commonly finds in studies across countries. So it emerges from your habilitation document.

Eva: It is getting old because I never find the time to . . . when switching contexts too often, one always happens to study something new and hardly finds the time to write things up properly. And each time I look at the data, I'm finding something new, of course.

Steve: You have been doing some work on the transition from school to universities, too, that is something that is still on going?

Eva: Yes, this is still going on, and it is also a project from Sweden. We, Christer [Bergsten] and I, looked into what has become called the transition problem. Our aim is to develop a more integrated view of this that takes into account social and pedagogical, or didactical, and mathematical issues. We have lots of data to analyze; we followed groups of engineering students who all have to take the same mathematics courses at maths departments in Sweden. We interviewed them in the beginning, in the middle and towards the end [of their first year of study at the university], and we looked at the teaching and the sort of maths and the exam results. We are beginning to see some interesting things, but it's quite complex because these different engineering programs are also recruiting different sorts of students. For example, in some environmental engineering, you would find more female students and they would be mostly middle class, intellectual middle class people; and in mechanical engineering, you would find male ones mostly and there would be more from working class contexts. This you might have guessed. As they are all confronted with the same university mathematics, we were *inter alia* interested in how their different careers and aspirations act out; how they see mathematics as helpful or not; how they would be able to recognize differences in knowledge criteria between this university math, which is quite Euclidean in the settings we study, in relation to their school math. We were also interested in how much they make use of provisions for helping them with mathematics, which are usually framed in a therapeutic way. We also asked them about their social activities: whether they feel they are in a group or whether they feel lonely, whether they are able to organize their own work, because a lot of the success in the first year just calls on their study habits, scholastic attitudes, how to keep up with the responsibility of looking after their own things. For example, some show the cleverness of getting old exam tasks and organize a working group. We want to integrate these different aspects into a broader view, rather than just looking at the problems they might face when dealing with mathematics; and we are beginning to find patterns there. We also found interesting views about how they think mathematics would be of use for their particular type of engineering professions [that] they are aspiring toward.

Steve: Does the way students use technology come into it as well; Wikipedia, and there are mathematics packages, and just googling things, does that come up?

Eva: Yes, this we have asked them in one of the interviews: What would you do if you feel you don't grasp some thing? They are looking up things in Wikipedia for example. But this depends on their social activities; those who are in groups prefer to ask other students and some ask teachers only.

Steve: We've talked about a few of the things you are involved in; and was there in separate projects, th[at] is, common themes coming through? In particular, you are clearly very interested in the whole socio-cultural perspective on what people are doing. I put it that way, because it's not uncommon to find researchers in our field saying: you know, this is the social context or this is the sociological stuff but I'm looking at the cognitive stuff. And you're looking at all these things in a much more integrated way. I mean, you're aware if you're zooming in, at the same time you're aware of the other factors that are around. And also your concern for who is succeeding and who isn't succeeding here; unpacking how that is coming about in different contexts. Could you say that?

Eva: Yes, I think that's the main interest, and I think it is obviously a major interest of all mathematics education researchers: to find some way of predicting, describing, or explaining, why students succeed and why they don't. And of course they do this from different political, theoretical, and methodological viewpoints. I think whatever the field of research is, it remains the main question in mathematics education. If research has a face towards practice, I think I would want to frame it as a specialization of a general theory of curriculum and pedagogy; which is perhaps a German tradition. We should also look into other school subjects and not always package mathematics and science. I would always have loved to see the students from learner's perspective study in a science lesson, in a music lesson, in religious education or in a language lesson. Because the field is so isolated, we don't even create such data. What you said was about looking from a broader angle and in an integrated way: [it] is also a troubling issue, because one tends to read all kinds of published research from all kinds of theoretical vantage points, for example when researching the so-called transition problem. But there can't be any consistent perspective through which you can re-read or recontextualize all other research [that] deals with a similar issue. This is troubling, because one choice is to just create your own project; develop your own consistent framing, basically by drawing on some specific set of theory, in parallel to other such projects. There is a serious problem with an integrative view, not only for an individual researcher. It's a problem for the whole community.

Steve: Hm. If I asked you to describe where interests and concerns (in particular with who succeeds and who doesn't), where that comes from, how far back would you need to go? I understand you started off by studying psychology and psychiatry, and then you moved into mathematics and philosophy; and you did some teaching and then you moved into math education. I'm interested in, you know, if you try and trace back; [is] there some route somewhere where your current interests come from.

Eva: I would at the very moment when trying to answer your question try to produce a consistent narrative, which I can't spontaneously produce; and I don't think that there is a way I would ever be able to do it, because I think there are more arbitrary situations one ends up in.

Steve: What about your first school teaching experiences; is that the way you encountered situations that were perhaps different from your own school even? When I was teaching I certainly remember, realizing some students came from such different backgrounds to me, and what did that have to do with the way that they were succeeding or not succeeding with mathematics, or how I was interacting with them . . . a host of questions for which I had no answers, but at least a growing concern that there was something here worth me thinking about.

Eva: I think it really does not come from my teaching experience, I think it comes from my experience as a school student. When I went to school in Austria, I remember very early experiences from my primary school; because when I moved context, I did not fully understand or speak the other kids' dialect. And tried to make up for this, by helping some of them with their homework. I remember on the way to school, it was in the countryside about 4 kilometers walk, we were meeting a little bit earlier and I wrote the answers to, for example the arithmetic tasks, in their notebooks. I was distributing my homework to those who couldn't do it, and I remember that these students were marginalized, as they were said to come from abnormal families. And I developed a solidarity with them because I also felt marginalized. I continued teaching them. This teaching experience went on, as I lived in a village where a lot of people had children just about three or four years younger than myself; their parents, who all had academic qualifications and so high aspirations for their children, wanted them to do well at school. I was teaching these children during my upper secondary career in almost all subjects. I think this is an experience that might have contributed to my views about teachers, about school, and who these people are who don't achieve for very different reasons. I also found school often quite awful; and I did not go to school much for some time in summers, because it was much more fun to swim in the lake nearby. It had to [deal] with the authority relationships in the school and with fairness. I think fairness was an important concept.

Steve: What brought you to initially psychology and psychiatry in university?

Eva: Psychology was part of a teacher-education program much later, but psychiatry was when I was interested in refractions perhaps; deviations, diversions from the normal. I think I have always been

interested in deviation. For some reasons, I knew people who
were in psychiatric wards; so there is some contact with deviating
people, which might create such an interest. It was always puz-
zling to me how explanations and theoretical framings of such
deviations operate. And who is framed as deviating, and who
isn't, and whether this is a social construction, or what of this
would not be accessible to being explained as a social construction, and what are the sorts of treatments invented. I worked in a
psychiatric ward as a summer job, and this was the point where I
switched to studying mathematics and philosophy. Because they
told me that at this point, when people are locked in for 10 or
15 years, there is nothing you can do about it. It's creating these
deviations earlier in the normal practices of societies, where the
point is. That made me switch to something else.

Steve: I remember texts. I never went into psychology or psychiatry,
but read quite a few texts in the 1960s and 1970s that gave different interpretations towards what schizophrenia was all about;
and it was about families and so on, but...So you turned into
mathematics and philosophy, right?

Eva: Hm.

Steve: And I hear stories about you serving in bars or cafés in the night
in places where philosophy was being discussed.

Eva: Yes, this was very nice; it is of course always this mythologizing of
the past, a *nostalgia*, but I seem to remember it as a very happy
time. This was during my undergraduate studies in Vienna,
which obviously took some time because I switched my subjects; and there I earned some of my money through work in a
café, which was at the same time a gallery. It wasn't really a very
beautiful place, as you might imagine a café in Vienna, but it
was at a corner where interesting people from Vienna dropped
in mostly between 1 and 3 o'clock in the morning. All kinds of
people, artists, local politicians, even some celebrities you would
occasionally see there. Also many fellow students came, philosophers, and mathematics colleagues. So we discussed in this
context about just everything with everybody.

Steve: Vienna is such an important city from the point of view of ideas:
Freud, Wittgenstein...

Eva: Yes (laughing)

Steve: Was Wittgenstein a person whose thinking came up a lot, particularly? I'm just interested in his work myself, and have been
for many decades.

Eva: Yes, at that time he did. Curiously, Wittgenstein was a primary
teacher in the village where my grandparents went to school;

but this I didn't know for a long time. What came up from Wittgenstein's work, was mostly the later work: *Philosophische Untersuchungen*. We also discussed Gödel's proof for quite some time, and we came to the conclusion that somehow it was a trick, it was wrong.

Steve: This would be at 3 in the morning...

Eva: Yeah (laughter), I still think it's wrong.

Steve: (laughter) Even at 2 in the afternoon.

Eva: (laughter) It was a lot about analytical philosophy of science, history of science and its framing; it was also about other subjects like writers, often Austrian writers. These discussions were usually initiated when any of the people, if they were colleagues from the university, had to write a seminar paper about something, so I think we have collectively written all kinds of essays, assignments, in this café during evenings; and everybody contributed, and we learned a lot because it was people who had to write about Robert Musil and Ingeborg Bachmann. And then people had to do something about Gödel, or about theory of evolution, and so there was a context where we engaged in each others' studies across all kinds of subjects.

Steve: Is that experience something you continue to draw on, interest in philosophy and the range of things that you were talking about there?

Eva: Yes, but I find it more and more irritating too... because somehow it is interesting to "draw on," and if one draws on very different ways of reading things and interpreting things and framing things, then there is a repertory [that] is extremely hard to bring into some consistency for yourself. I'm not so sure whether I understand myself how this "drawing on" really works. It seems to work in a very associative way. But surely I draw especially on the readings, because it is a lot of time we spent reading a lot of things and you start knowing a lot of things.

Steve: That brings me to what I was thinking about as perhaps two winding up questions. Perhaps this drawing on, we could start there. I was going to ask you two questions to end up with. What are the things in the field, in our work as researchers in the field, that concerns you most, bugs you most, irritates you most; and then to end positively, what are the things that you are enjoying seeing developed within our field, which you may or may not be a part of. Starting off with the things that bug you, the things that annoy you, I wonder whether the way that many of us talk about "drawing on" theories, and what we actually are doing, is one of them or not.

Eva: Without much thinking, there is some notion of whether you find it interesting or not. And it seems to be, for me at least, quite independent of what theory or framework they are drawing on; or whether they do draw on any framework at all. It's perhaps [has] to do more with consistency, and whether it's making a point; which I haven't thought about in the particular way before. The troubling thing is that there are not many such things to read. One can't systematically read things any more, because there are so many things published; but one reads a kind of arbitrary selection, possibly one searches for some journal article and sees titles of five others, and looks.... But the troubling thing is that it doesn't happen frequently; that one is drawn into something which creates such an interest through a particular way of describing things. I think it has to do with the tremendous fragmentation of the field. What is troubling, is the feeling I get when I read things that are completely obvious already; or one has already seen it before, in many slightly different versions, and it seems to be quite clear that the people who wrote it don't seem to be aware of this fact because there are always about 15 key references that pop up in my head when I read a paper, that are omitted. For example, I remember that the same issue was already discussed by an old German didactician, or in the 1970s there were some French who already solved this problem. It's interesting because people seem to have started to forget a whole repertory of research in mathematics education that was produced in its beginning, when the first generation of inventive people created this field. And it started to get standardized, and differentiated and fragmented, so that one feels a little bit lost. There's a problem in this fragmentation, and in the quality of research outputs very often. I think this is particularly interesting.

Steve: You and I worked together on the *Encyclopedia of Mathematics Education*, which is due out quite soon, and we as a group of editors all tried to ensure that people writing entries looked at where those ideas, that they were writing about, actually came from. My hope is that that's going to lead to people becoming more aware of what's gone before when they get hold of, I don't know constructivism or whatever, to see where these ideas came from and perhaps not forget all about it and start more recently, but to draw things together. I don't know, do you think that we might have played a small part in that?

Eva: Well I am not sure how many are reading *Encyclopedia* articles and research syntheses, especially not if they come in unaffordable expensive publications. I'm afraid I'm somewhat pessimistic about the impact of this encyclopedia; but I think that the math education entries are much better than articles in many similar encyclopedias. If we see the research as a cumulative evolutionary activity, then the worry is that we are reinventing a wheel all the time. But if research is about publishing, then I think there is a need to reinvent the wheel all the time.

Steve: So what about looking more . . . what things do interest you and excite you about things going on right now, I hope there are some at least (laughter) . . .

Eva: What I find exciting is a new real broad range of views or framings of the empirical that's developed. I'm also fascinated by the knowledge of some people working within mathematics education, who seem to be able to better summarize some works of Lacan or Foucault or other hard stuff than some people who write introductions into sociology. Obviously some mathematics education researchers are very good in using frameworks, and rewriting them. I find it fascinating, if I see people who pursue such projects consistently. This is done from within institutions where the people are professors in mathematics education and not professors of something else. I think this is a very good thing to see; and I think there needs to be some strategy of how one deals with this, because it's about the identity of the field as such. I think it might not be noticed by people who are, well, generic professors of maybe linguistics or sociology, who might not even be aware of the recontextualization of the knowledge from their fields that people are creating in mathematics education. I think it is interesting what constitutes expertise, then, in different academic fields. This development is really interesting, especially how far some people get with this and how well they are able to do it. People need to find some time to engage with these things.

Steve: Let us stop there, thank you.

Eva: Thank you, Steve.

Steve Lerman
South Bank University, London
United Kingdom

CHAPTER 8

MATHEMATICS
AND PHILOSOPHY
A Semiotic and Historical Perspective

Michael Otte

INTRODUCTION

The importance of philosophy to mathematics education results not least from the fact that the concept of "explanation" is central to our educational practices and aims. Without content, mathematics could not fruitfully be organized and pursued at school as a primarily professional topic or as a mere language. Mathematical education has, like other subjects, also contributed to a common search for clarity on fundamental issues and to the formation of the person; such that philosophical formation seems as important as logical exactness or mathematical literacy.

When Cardinal Bellarmino notified Galileo of a forthcoming decree of the church, condemning the Copernican doctrine of helocentrism and ordering him to abandon it as an explanation of the world, he used an argument saying that mathematicians always used to argue hypothetically or *ex suppositione* only:

> First, I say it seems to me that your Reverence and Signor Galileo act prudently when you content yourselves with speaking hypothetically and not

Refractions of Mathematics Education, pages 131–154
Copyright © 2015 by Information Age Publishing
All rights of reproduction in any form reserved.

absolutely, . . . For to say that the assumptions that the Earth moves and the Sun stands still saves all the celestial appearances better than do eccentrics and epicycles is to speak with excellent good sense. . . . Such a manner of speaking suffices for a mathematician. . . . To demonstrate that the appearances are saved by assuming the sun at the centre and the earth in the heavens is not the same thing as to demonstrate that *in fact* the sun is in the centre and the earth is in the heavens. (Bellarmino, 1615)

As logic or mathematics can only define concepts, rather than objects, Bellarmino seems right. Galileo himself says, for example in his *Esercitazioni filosofiche di Antonio Rocco* or at the end of the 3. Day in the *Discorsi* that he has reached certain convictions about the fall of bodies and the involved velocities by thought experiments and *before* experiments or observations had taught him (Enriques, 1927, pp. 51ff.; Weyl, 1966, p. 200).

Scientific explanation is judged from a certain perspective or world-view, and in the context of a certain language. Speech seems the essential connective between subject and object. To make something the object of investigation one has to form a related concept. The essence of something is from now on to be grasped within the logic of its further rational development and in the possibilities thereby opened up.

Mathematics, in particular, being by its very nature exclusively concerned with meta-data, which it produces sometimes itself, rather than objects, has to uncover the hidden logic of our stories about the world (Paulos, 1998). Our stories, in return, have to give meaning to the mathematical diagrams, calculations, and deductions.[1]

Now modern societies presuppose first of all regular and standardized ways of organizing both our concepts and our institutions. The explanatory schemata resulting from this standardization, tends to destroy individualism and enchantment. But mathematics education is in fact the only place to treat the human subject's relation with mathematics. And that is what mathematics education is all about: make the human subject grow intellectually and as a person by means of mathematics. In this chapter we will discuss the relationship between philosophy and mathematics from the point of view of mathematization and with reference to the historical development, because we believe that the relationship between mathematics and philosophy becomes best accessible in this way.

CONCEPTS AND OBJECTS, OR THE COMPLEMENTARITY OF INTENSIONS AND EXTENSIONS

Plato's philosophy arose from a scandal, namely Socrates condemnation and death in 399 BC. Socrates had obviously been a virtuous and wise man; how could that have remained hidden from the people of Athens? And how could they be brought to recognize or accept truth? Plato blamed the Sophists for this scandal, and in his dialogue *Sophist*[2] he emphasized that there were different kinds of speech and consequently different forms of human existence; not all served to disclose the truth. To this purpose he identified two kinds of activities or arts: productive and acquisitive arts (219, c-d) and two kinds of languages, those which related to real things and those which referred to other signs (265, b-c). He finally defined the Sophist as somebody who claims to be able to speak about anything without real knowledge, and who forces the person who converses with him to contradict himself (268, b). The interesting thing is the claim that different kinds of languages or signs lead to different kinds of human existence.

In Plato's dialogue *Cratylus* two primary interlocutors of Socrates, Hermogenes and Cratylus, represent two diametrically opposed views on the status of language, and in particular two opposing answers to the question about the origin of words (nouns, names, etc.). The positions of Hermogenes and Cratylus have come to be known as conventionalism and naturalism (or essentialism), respectively. In the dialogue, Socrates was asked whether language is a system of arbitrary signs or whether words have an intrinsic relation to the things they signify. Plato's essential interest was to see whether language served primarily communicative and rhetorical purposes, as the Sophists believed and tried to argue; or whether language was essential to cognition and objective knowledge.

An extreme linguistic conventionalist like Hermogenes holds that *nothing* but local or national convention determined which words were used to designate which objects. Cratylus, as an extreme linguistic naturalist, holds that names cannot be arbitrarily chosen in the way that conventionalism describes or advocates, because names or concepts belong *naturally* to their specific objects. The Sophists (Protagoras) were described as conventionalists while Plato's rather anti-conventionalist views lead to the conviction that things have objective natures independent of how they may appear to us, and that there are objectively determined skills for dealing with them.

Socrates forced Hermogenes to admit that any purposeful activity—even the efforts of a Rhetorician or of a straightforward liar—is objectively constrained, if it wants to be successful. And this is the interesting aspect of the story. It is the conditions of activity, rather than the objects as such and in themselves, which suggest the objectivity of reference. Mind and world are mediated or connected by the system of activities (including its means and goals). It follows from this in particular, that words or signs (on the one side) and objects and goals (on the other), are not as distinct and separated as one might suppose. To draw an absolute distinction between signs and objects, or between concepts and perceptions, the operative and receptive sides of the human mind, would amount to something like Xenon's paradox of the race between Achilles and the Tortoise (Peirce, CP. 5.157 and 5.181).

Let us have a look at Plato's Dialogue *Cratylus:*

> **Socrates:** Does what I am saying apply only to the things themselves, or equally to the actions which proceed from them? Are not actions also a class of being?
>
> **Hermogenes:** Yes, the actions are real as well as the things.
>
> **Socrates:** Then the actions also are done according to their proper nature, and not according to our opinion of them? In cutting, for example, we do not cut as we please, and with any chance instrument, but we cut with the proper instrument only, and according to the natural process of cutting; and the natural process is right and will succeed, but any other will fail and be of no use at all.
>
> **Hermogenes:** I should say the natural way is the right way...
>
> **Socrates:** And this holds good of all actions?
>
> **Hermogenes:** Yes.
>
> **Socrates:** And speech is a kind of action?
>
> **Hermogenes:** True.
>
> **Socrates:** And will a man speak correctly who speaks as he pleases? Will not the successful speaker rather be he who speaks in the natural way of speaking, and as things ought to be spoken, and with the natural instrument? Any other mode of speaking will result in error and failure.
>
> **Hermogenes:** I quite agree with you.
>
> (Cratylus, 387 a-e)[3]

A sign, says Charles Peirce:

> Is something which stands to somebody for something in some respect or capacity. It addresses somebody, that is, creates in the mind of that person an equivalent sign, or perhaps a more developed sign. That sign which it creates I call the *Interpretant* of the first sign. The sign stands for something, its *Object.* It stands for that object, not in all respects, but in reference to a sort of *Idea,* which I have sometimes called the ground of the representamen. *Idea* is here to be understood in a sort of Platonic sense, very familiar in everyday talk. (Peirce, CP 2.228, CP 2.275, CP 5.283)

Peirce sees the sign as primarily determined by its object, while conceiving of the distinctions between object, sign and interpretation of a sign as being only relative; and classifies the signs accordingly, in a first access, with respect to the nature of their objective reference into three classes: *indices, icons* and *symbols.* Indexical signs are essential, because they ensure objectivity, while iconic characters are indispensable in the genesis of new ideas. Conventional signs, Peirce calls them symbols, represent a kind of synthesis of icons and indices. A sentence would be for Peirce an example of a symbol; the subject of the sentence represented by an index and the predicate by an icon.

As indices represent the referential and icons the intentional aspect of a sign, one might conclude that knowledge development is governed by the complementarity of intentional and extensional aspects of signs. Alternatively, all discovery or creation is a combination of chance and intuition, of hitting more or less accidently on some object and interpreting it metaphorically. A very nice example of Freudenthal is discussed in Otte and Zawadovski (1985).

Peirce, in fact, defines all cognition or creativity in semiotic terms, that is, as *semiosis*, defining semiosis as the action or process of a sign. "By 'semiosis,' I mean," Peirce writes, "an action, or influence, which is, or involves, a cooperation of three subjects, such as a sign, its object, and its interpretant; this tri-relative influence not being in any way resolvable into actions between pairs" (Peirce, CP 5.484). Representation, in the sense of Peirce, thus is triadic, rather than dyadic.

Niels Bohr's term *complementarity* has been used by a number of authors to capture the essential aspects of the cognitive and epistemological development of scientific concepts. In this chapter we will conceive of complementarity in terms of the dual notions of extension and intension of (mathematical) signs (Otte, 2003). A complementarist approach is induced by the impossibility to define mathematical or scientific reality independent from activity conceived of as a mediating instance.

DICHOTOMIES DURING THE "CLASSICAL AGE" (FOUCAULT)

In his *The Order of Things*, Foucault (1973) has described the cultural transition, which characterized the *Scientific Revolution* of the 16th and 17th centuries in terms of a transition from the *Age of Interpretation* to the *Age of Representation*; that is, from a view of signs in the sense of Cratylus' naturalism to linguistic conventionalism and philosophical nominalism. According to Foucault, in its most originary form, language was thought of as a certain and transparent signature of Nature, due to immediate resemblance with designated things. In the Renaissance, the episteme of similitudes still prevailed to a certain extent, and thus direct interpretation had the power to reveal the true nature of things. During the period of the Scientific Revolution, things changed and the episteme of representation and of linguistic activity underlay all knowledge; that is, symbols became classified according to their sense, that is, according to relations of identity and difference, rather than according to objective reference.

Foucault says:

> At the beginning of the seventeenth century during the period that has been termed, rightly or wrongly the Baroque, thought ceases to move in the element of resemblance. Similitude is no longer the form of knowledge, but rather the occasion of error.... Resemblance, which had for long been the fundamental category of knowledge...became dissociated in an analysis based on the terms of identity and difference. (Foucault, 1973, pp. 51, 54)

Foucault denies the special role of mechanics and mathematics in the Scientific Revolution and since (p. 57). But mathematics is pre-eminently the science established by specifying things simply as being equivalent or different, without assigning a real content to these relations.

The topologist Salomon Bochner considers the iteration of abstraction as the distinctive feature of the mathematics since the Scientific Revolution of the 17th century.

> In Greek mathematics, whatever its originality and reputation, symbolization...did not advance beyond a first stage, namely, beyond the process of idealization, which is a process of abstraction from direct actuality,...However...full-scale symbolization is much more than mere idealization. It involves, in particular, untrammeled escalation of abstraction, that is, abstraction from abstraction, abstraction from abstraction from abstraction, and so forth; and, all importantly, the general abstract objects thus arising, if viewed as instances of symbols, must be eligible for the exercise of certain productive manipulations and operations, if they are mathematically meaningful....On the face of it, modern mathematics, that is, mathematics of the 16th century and after, began to undertake abstractions from possibility only in the 19th century; but effectively it did so from the outset. (Bochner, 1966, pp. 18, 57)

In the 19th century, only mathematics proceeded from problem solving, and analysis to construction, and experimentation with theories (Leibniz might be an exemption).

Through naming and conceiving things, Foucault continues, the being of things was established and a universal science of order was developed. And all the classical sciences were nothing but "well-formed" languages. Nomination was a key, because in the classical period, language was a main form of knowing.

> And knowing was automatically discourse.... It was only by the medium of language that the things of the world could be known. Not because it was a part of the world, ontologically interwoven with it (as in the Renaissance), but because it was the first sketch of an order in representations of the

world.... Classical knowledge was profoundly nominalist. (Foucault, 1973, pp. 295–296)

To know from now one means to present, rather than to interpret; and linguistic representation played an eminent role alongside activity and construction. Language represented thought and thus ordered it. Foucault views appear one-sided, however, as he fails to mention that the establishment of activity (experimentation, construction, navigation, printing and technology in general), became fundamentally important, alongside language and rhetoric, as an institution mediating between the epistemic subject and the objective world. Already in 1620, Francis Bacon had said that printing, gunpowder, and the compass were the three inventions that "have changed the appearance and state of the whole world." No social theory of knowledge can be developed ignoring the role of technology. Real knowledge requires a harmony between mind and nature, to be realized within the bounds of human activity and practice (Otte, 1993).

The epistemic role of activity explains or helps to explain the analytic character of modern science, as well as its rational, anti-Aristotelian character. Descartes indicates the new rationalism by saying in his *Discours de la Methode* that the effects, which are to be conceived of as the consequences of fertile hypotheses, have to establish the causes, rather than the other way around:

> For since experience renders the majority of these effects most certain, the causes from which I deduce them do not serve so much to establish their reality as to explain their existence; but on the contrary the reality of the causes is established by the reality of the effects. (Descartes, orig. 1637, Part VI)

This reflects the analytical character of the classical sciences, while the impact of empirical data and related processes of synthesis came somewhat later to the foreground. And hence results the central epistemological importance of the notion of complementarity, although its importance came to be recognized profoundly during the 19th century only. During the 17th and 18th centuries, styles of presentation still remained largely unconnected or even in opposition to each other.

Consider geometry as an example: Euclidean geometry in antiquity was a geometry of figures, not a theory of space. There was no global space concept, and this caused confusions, which Xenon expressed in his paradoxes. In the 17th century, geometrical thinking of space developed in two rather different directions and according to two different styles of thinking, represented by the names of Descartes and Desargues. The infinite straight line of Cartesian geometry is something like a measuring rod; whereas in the Desarguean context, it is conceived of as the ray of light or perception (Otte, 1991).

This dichotomy became more profound during the 18th century. Mathematics became troubled by a certain rift between positivistic reductionism

(Cartesianism), and the desire to explain advanced mathematics by elementary mathematics, on the one side; and formal algebraic analysis, on the other side. In particular "the major philosopher of method in France, Etienne de Condillac, carried the eighteenth-century enthusiasm for analysis to an extreme" (Hankins, 1985, p. 21). Algebra, says Condillac, for example, is a perfect analytical language. Condillac also stresses the social nature of signs and of semiotic activity: "Before social life, natural signs are properly speaking not signs" (Condillac, 2001, p. xxvii). But above all, Condillac's semiotic conceptions were exemplary, in as much as they reflect the task of the sciences of the Classical Age (Foucault) to reconcile or combine the mathematical approach with a thorough sensualism and empirism. It is the notion of (semiotic) *activity* that is crucial.

MIRACULOUS NUMBERS AND A NEW PHILOSOPHY

Each measurement positions the measured (the measured value within an arithmetic algebraic structure), providing it with, at times, mystical meanings. A thing or a fact has no meaning; it must become a sign to get a meaning. It is the symbols and representations and their mysterious internal relationships or structures, and the apparent life of its own that they seem to have, that gives knowledge (or even our experiences) their mystery, their attraction and charm.

8, 105, 121: these strange numbers represent more than anything else the attraction of mechanical philosophy and the mystery of the mathematics of the universe. The 16th and 17th centuries modeled nature on the characteristics of a machine; that is, in terms of a huge clockwork and in contrast to Aristotelian animism. Our language of ideas up to now owes a great debt to that mechanical model of the solar system.

Edmond Halley, Newton's friend, convinced the latter to write the *Philosophiæ Naturalis Principia Mathematica* (1687), which was published at Halley's expense, and which made Newton the greatest scientist of modernity. Newton's *Principia* marks, conceptually, a radical departure from the then-dominant tradition of Aristotelian science. Aristotelian science of the empirical phenomena was descriptive and qualitative. With Newton, it was to explain nature in mathematical terms, rather than speculating about the essence of things.

And it was Halley who wrote in 1716 an

> essay that called upon scientists to unite in a project spanning the entire globe; one that would change the world of science forever. On June 1761, Halley predicted, Venus would traverse the face of the sun. (Wulf, 2012, p. xxiii).

He believed that measuring the exact time and duration on different locations on the earth, would enable scientists to calculate the distance between earth and sun. The dates of the Venus transit: 1761, 1769, 1874, 1882, 2004, 2012. So the exact numbers are: 8, 105.5, and 121.5 years. Halley's request would be answered when hundreds of astronomers, sailors, soldiers, statesmen, instrument makers, and whoever else joined in the transit project. The cooperative efforts to observe and measure the Venus transit, was a pivotal moment in the new era; an epoch in which man tried to master nature through the application of mathematical principles, rather than relying on metaphysics and historical or religious narrative. Measure and structure became means to organize knowledge and to make new discoveries.

God had made "the world according to weight and measure," we read in Diderot's great *Encyclopédie* and we humans are small copies of him. Although God is the only one to have a complete mathematical and philosophical knowledge of it. Our measuring is only approximate and has needs for tremendous theoretical efforts. Instead of grasping the exact value for π for example, we must be content with inexact values derived from some similarity between quadrangle and circle. Leibniz did frame his idea of the "complete concept of an individual substance" from such examples; and he held the belief that there is a gradual difference only between knowledge and reality, due to our limited capacities.

THE ROLE OF MATHEMATICS IN THE RISE OF SCIENCE

Galileo claimed that the *Great Book of Nature* was written in mathematical symbols;[4] geometric figures and his words have in the centuries since manifested their truth beyond any measure that Galileo could have imagined. Galileo also stressed that the book of nature, "cannot be understood unless one first learns to understand the (mathematical) language and knows the characters; without these it is humanly impossible to understand a single word of it." And this emphasis on mathematics was essential to Galileo, as various of his letters and writings show (see for example his letter of 1615 to the Grand Duchess Christina), because he wanted to avoid conflicts with the words of the holy scripture and emphasized that nature does not speak by itself (McMullin, 1998).

The new science became acceptable only after that transition about which Foucault speaks; that is, after it became acceptable or even obvious that there were no "literal" meanings or truths and Galileo tried to argue this. Theories relate directly to theories only; works of art refer primarily to works of art.

Why was mathematics and its special language so important to the new concept of exact sciences of 17th and 18th centuries? Have not quite a number of scholars (Foucault, Alexander etc.) even claimed that mathematics was exempt from the great discoveries that characterized the Scientific

Revolution. Often mathematics is, in fact, simply identified with scientific certainty; this to the expense of its relevance for the experimental sciences. Alexander underscored this view in his statement about the relation between experimental science and mathematics:

> While the experimental philosophers could easily imagine themselves as explorers of the secrets of nature, the case was more difficult for mathematicians. Mathematics, with its rigorous, formal, and deductive structure, appeared to be an ill-suited terrain for intellectual exploration.... Mathematicians, it seemed, did not seek out new knowledge or uncover hidden truths in the manner of geographical explorers. Instead, taking Euclidean geometry as their model, they sought to draw true and necessary conclusions from a set of simple assumptions. The strength of mathematics lay in the certainty of its demonstrations and the incontrovertible truth of its claims, not in uncovering new and veiled secrets. (Alexander, 2001, p. 2)

Contrasting exploration with mathematics, and making that the basic background philosophy might seem rather hasty or even wrongheaded. One should perhaps interpret Alexander's observations in a radically different way and consider mathematics from the point of view of mathematization. Mathematics and logic are concerned with relations between facts or informations, with *meta-data*. This train was important, because science does not consist in collecting individual data. Its special power is derived from theories, from the structured arrangement of data and from the consequences, which can be drawn from it. The mathematical deductive method has given physics its predictive power, and Newton's discovery is a compelling illustration of this fact.

When one speaks of empirical science, one must not forget that observation and experiment were only able to establish modern science, because they could rely on mathematical deductions (Reichenbach, 1951). And Salomon Bochner introduced the following comparison between empirical science and mathematics:

> If science is viewed as an industrial establishment, then mathematics is an associated power plant which feeds a certain kind of indispensable energy into the establishment. (Bochner, 1966, p. 47)

Bochner's comparison hinted at the double role of symbolization. Any sign has a sense and a reference. Sense is derived from the internal relationships of the symbol system or language as a whole. The sense of an algebraic equation lies in the possibilities of calculation, including its possible solutions; the references depend on the intended applications or models. In this way the evolution of mathematics becomes governed by the complementarity of the intensions (meanings, senses) and extensions (references) of mathematical signs.

The concept of a complex number, for example, is an abstraction from possibility, "inasmuch as it is possible to extend to such objects the basic arithmetical operations of addition, subtraction, multiplication and division, in a suitable manner" (Bochner, 1966, p. 55) and thereby solve equations, which otherwise were not solvable and thus would not "make sense." Does the real number x, which makes the equation $x^2 = -1$, or written differently, $x = -\frac{1}{x}$ true, exist? If so, it must be equal to 1 or to −1, and this yielded $1 = -1$; a contradiction. But the mathematician enlarges his universe and finds a new system of numbers; complex numbers and thereby an enlarged set of roots of unity.

As long as the imaginary numbers had gained admission to arithmetic as a calculatory symbol only, their use produced the most horrible confusions. It was not even clear what a simple equation $A = B$ could mean and how it should be handled (cf. Nahin, 1998). Only after Gauss had given a relational interpretation to the imaginary unit in the frame of the model of the so-called Gaussian number-plane, it became a legitimate mathematical object. This subsequently assumed an important role in function theory during the 19th and 20th centuries. This metaphysics of imaginary numbers is based on two assumptions: namely that they are subject to all arithmetical operations, and further that we can form an intuition of their objective meaning.

Once more we come to the conclusion that the complementarity of intentional and referential use of concepts provides an essential orientation, as well as a fundamental problem.

A COPERNICAN REVOLUTION IN PHILOSOPHY, WHICH WAS NOT COMPLETELY SUCCESSFUL

The Copernican revolution forever changed the place of man in the cosmos, and thereby changed the idea of what it meant to explain that cosmos. And modern philosophy or metaphysics, says Burtt, "is in large part a series of unsuccessful protests against this new view of the relation of man to nature." Berkeley, Kant, Fichte, Hegel—all are "united in one earnest attempt, the attempt to reinstate man with his high spiritual claims in a place of importance in the cosmic scheme" (Burtt, 2003, p. 25).

Kant, for example, set out to complete Newton's scientific revolution (Hahn, 1988). He believed that the idea of knowledge did not lie in the object as such, but was rather based on the conception of the (epistemic) subject; transferring the subject-object distinction to the plane of our cognition. "Our knowledge springs from two main sources of knowledge," concepts and intuitions (Kant, 1787/1990, p. 74). The interplay between these two faculties realized itself within the system of human activity. Kant's concept of the human subject remains, however, rather static, being confined within the structures of the transcendental subject, which seems to replace the god of classical rationalism. There was no more a privileged or God's view of subjectivity than there was an objective knowledge of the thing in itself.

Kant had conceived of the difference between mathematics and philosophy in terms of method, rather than in terms of subject matter, and it came down to a description in terms of what we have called complementarity. He writes: "Philosophical cognition is rational cognition by means of concepts, mathematical cognition is cognition by means of the construction of concepts (in intuition)" (Kant, 1787/1990, p. 741).

Mathematics does not proceed from concepts alone, but rather has to rely on particular instantiations of these, as well; that is, on particular objects, carrying out arguments in terms of such particular representatives, arguments which cannot be carried out by the sole means of general concepts. Even in a deductive proof one might argue, for example, that line A is parallel to line B, or intersects with it at point C. Kant took great pains to draw our attention to these aspects of proving. There is no activity without an object. Mathematics cannot establish its objects by means of description after the fashion of Leibniz' notion of the "complete concept" of an individual substance. Classical thought rested in the idea of God. The proof of the existence of God warranted Leibniz' foundation of truth on proof, as well as the Cartesian cogito ergo sum; this final truth constituted the foundations of the entire structure of Cartesian rationality. Kant takes great pains to distinguish analytic and synthetic propositions, because his view of the analytic-synthetic distinction depends on the invalidation of the ontological proof of God's existence, and it represents his own Copernican step.

Kant's answer to his fundamental question—how are mathematics and exact science possible?—in terms of the structure of a universal epistemic subject, was a step forward. But his conception of the human subject remained static, and a-historical and did not do justice to the dynamics of socio-cultural history and modern pluralism. Kant's essential problems are nevertheless preserved in various forms in the debates of logical positivism (Vienna Circle Philosophy), in pragmatism (Peirce, Dewey) and in Marxism.

THE SUBJECTIVE PERSPECTIVE: MENTALISM AND SOCIAL CONSTRAINTS

The Enlightenment was followed by the protests of the romantics, and the belief in a universal mathematical philosophy was lost.

The dominant focus of concern shifted away from the "interface problems" between individual cognition and the external world, and moved toward the problem of communication and the internal dynamics of society and culture. Mathematics turned into a social institution, and mathematical proof became established as its vehicle rather than as something merely serving individual conviction and certainty. Semantics and logics replaced epistemological concerns. Cavaillès comments on the situation as follows:

> Deux possibilités sont cependant ouvertes pour la doctrine de la science après l'analyse kantienne: suivant que l'accent est mis sur la notion de système démonstratif ou sur celle d'organon mathématique. (Cavaillès, 1976, p. 14)

J. J. Rousseau, sometimes called "the patron of the romantics," was perhaps the first person who searched for the "human soul" as a fountain of inspiration and scholarship. Kant had called him his "second Newton." And the great anthropologist Claude Lévi-Strauss, who was convinced that Rousseau had articulated the fundamental anthropological problem of the relationship between nature and culture, and that it had thus enabled us to speak fruitfully about the human subject, writes in his *Tristes Tropiques:*

> Rousseau, of all the *philosophes,* came nearest to being an anthropologist. He never travelled in distant countries, certainly; but this documentation was as complete as it could be at that time ... Rousseau is our master and our brother ... and every page of this book could have been dedicated to him, had the object thus proffered not been unworthy of his great memory. For there is only one way in which we can escape the contradiction inherent in the notion of position of the anthropologist, and that is by reformulating, on our own account, the intellectual procedures which allowed Rousseau to move forward from the ruins left by the *Discours sur l'Origine de l'Inegalité* to the ample design of the *Social Contract,* of which *Emile* reveals the secret ... Never did Rousseau make Diderot's mistake—that of exalting the "natural Man." There is no risk of his confusing the state of Nature with the state of Society. (Lévi-Strauss, 1961, p. 389, emphasis in original)

Rousseau stands in opposition to the encyclopedists, especially to Voltaire, who, like Diderot, insinuated that Rousseau had favored a primitive reductionism. D'Alembert in his introduction to the *Encyclopédie* draws an interesting analogy with respect to Rousseau's *Discourse on the Moral Effects of the Arts and Sciences* (1750), saying that it might be interpreted to suggest that norms or laws contributed to the spread of vice. This is not as absurd as it might appear, given the fact that definitions and linguistic conventions narrow down reality and subsume it to certain perspectives and interests (remember the Cratylus dialogue). Rousseau, in fact, propagated a certain type of systems thinking, which another anthropologist Georg Forster, Alexander Humboldt's teacher, expressed some 30 years later. Forster always points out the systemic character of reality. A "negro," Forster says for instance was, properly speaking...

> a true Negro only in his own fatherland. Any creature of Nature is what it should be only in the locality for which it has been created; a truth, which is seen, confirmed every day in menageries and botanical gardens. A Negro born in Europe is like a greenhouse plant, a modified creature, in all properties subject to change more or less unlike that which would have become of him in his own fatherland. (Forster, 1785/1983, Vol. I, p. 13)

And with respect to the changes of scientific methodology, he criticized Kant's rationalism:

> As long as our insight remains limited we would seem far from an infallibility of principles. Will categorizations that are based on limited experience, while possibly useful within these limits, not appear one-sided and half-true once the horizon is expanded, the point of view displaced?...Perhaps our present scheme of the sciences will become obsolete and deficient half a century from now, just like the previous ones. (Forster, 1785/1983, pp. 5–6)

Rousseau, in a similar methodological spirit, wrote at the beginning of his *Discourse upon the Origin and the Foundation of the Inequality among Mankind:*

> Let us begin by laying aside facts, for they do not affect the question. The researches, in which we may engage on this occasion, are not to be taken for historical truths, but merely as hypothetical and conditional reasonings, fitter to illustrate the nature of things, than to show their true origin, like those systems, which our naturalists daily make of the formation of the world.

Reminding ourselves of the discussions between Bellarmino and Galilei, it might appear that in the beginning moments of a new science and a new field of study, preconceived ideas and assumed hypotheses play a more or less fundamental role. The romantics were concerned with the human subject, and the subject could not be defined and replaced by a model or description of it. Although we cannot otherwise theorize about it. With respect to mathematics, these developments led into an antagonism between intuitionism and formalism.

PURE MATHEMATICS AND THE INDUSTRIAL REVOLUTION

Since the Industrial Revolution, our societies have become pluralistic and individualistic in every aspect and all districts of life. These developments have brought about, on the one side, a new interest in theory; and in formal axiomatic and rigorous proof procedures, on the other side. In the course of this second movement, 18th century algebraic analysis in the sense of Euler and Lagrange, was replaced by arithmetized analysis and function theory in the sense of Bolzano and Cauchy. The arithmetization of mathematics since Descartes, Bolzano, Cauchy, and Weierstrass, became decisive to the 19th century educational context and the development of pure mathematics. A first expression of its reductionism becomes visible from the manner in which the Bolzano-Cauchy approach dealt with the problem of continuity and of the continuum.

One of the first observations of the definition of continuity in the Bolzano-Cauchy manner, concerns its "local" character (in contrast to the traditional concept of "uniform continuity"); the definition speaks about continuity at a certain point and then might generalize by quantifying over point sets. This understanding of the continuum in terms of point sets, eliminates already all real continuity and generality. The only requisite for the definitions of and theorems on continuity is the availability of a notion of *distance* as a measure of *proximity*. This leads to the abstract notion of a *metric space*. A slightly deeper analysis of the relationship between a given metric on a set, and the collection of functions continuous with respect to that metric, shows that it is not the metric that is significant; but only those subsets which are *open*, where this notion of *openness* is defined as a generalization of the notion of an open interval of the real line.

Thus the task of characterizing continuous functions is equivalent to choosing a topology, that is, choosing a class of open sets. And if we choose the discrete topology calling all individual points of the line "open," all functions are continuous. The concept of continuity becomes formal and, in a sense, empty.

There have been, however, as was just said, two different trends in the foundational debate of mathematics during the 19th century, for which the contrasting conceptions of the continuity principle of Cauchy and Poncelet mark a significant expression (Belhoste, 1981; Israel, 1981; Otte, 1989). Rather than conceiving of this principle in terms of structure and variation or invariance, Cauchy thought of continuity in terms of approximation and limit.

The axiomatic movement, in contrast, which was anticipated in the work of Poncelet or Grassmann, tried to employ a top-down strategy; solving the foundational problems of mathematics by extending and generalizing its relational structures and its rules of inference. Grassmann's dropping of the commutativity of a general product and his definition of the anti-commutative vector product provide a pertinent example here.

Axiomatic thinking is thinking about form, and form must be constructed and idealized. Whereas the rigor movement tries to synthesize from pre-given primary elements, according to narrowly fixed methods, the top-down strategy observes the indications emerging from the overall structure and its behavior under possible variation. The statements of Euclid could, for example, be interpreted in two complementary ways. Under one interpretation, the statement to be proved refers to a definite totality and it says something about each one of them. This is the usual understanding of set-theoretical mathematics. Under the other interpretation, no such totality is supposed and the sentence has much more conditional character and bore a strong analogy to a thought experiment.

> The major obstacle to an acceptance of the interpretation of Euclid's arguments as thought experiments is the belief that such arguments cannot be conclusive proofs. In particular, one might ask how consideration of a single object can establish a general assertion about all objects of a given kind. Part of the difficulty is due, I think, to failure to distinguish two ways of interpreting general statements like "All isosceles triangles have their bases angles equal." Under one interpretation the statement refers to a definite totality [...] and it says something about each one of them. Under the other interpretation no such definite totality is presupposed, and the sentence has much more conditional character—"If a triangle is isosceles, its two base angles are equal." A person who interprets a generalization in the second way may hold that the phrase 'the class of isosceles triangles' is meaningless because the number of isosceles triangles is absolutely indeterminate. (Mueller, 1969, pp. 299–300)

This last view is characteristic of the axiomatic approach. Here we envisage ideas of triangles in general, but they turn out to be ideas of particular triangles that are put to a certain use. On such an account, a general triangle is a free variable and not a collection of determinate triangles. It is an idea, which governs and produces its particular instantiations. Which properties are essential to a general triangle, depends on context, on the activity, and its goals. If the task, for instance, is to prove the theorem that the medians of a triangle intersect in one point, the triangle on which the proof is to be based can be assumed to be equilateral, without loss of generality; because the theorem in this case is a theorem of affine geometry and any triangle is equivalent to an equilateral triangle under affine transformations. This fact considerably facilitates conducting the proof, because of this kind of triangle's high symmetry.

Historically speaking Euclid's *Elements*, of course, cannot be interpreted as a formal system in Hilbert's sense. Mueller's description is nevertheless pertinent and illuminating, because the "hypotheses" or assumed premises are here to be seen as admissible or licensed constructions, rather than propositions. In Euclid's geometry, the stipulated possibility of performing a certain construction, leads as a consequence to the possibility of other constructions; and this possibility is to be found out by a thought experiment. Descartes and Newton understood Euclid in this manner. Newton, for example, wrote in the preface to the first edition of his *Principia*:

> The description of right lines and circles, upon which geometry is founded, belongs to mechanics. Geometry does not teach us to draw these lines, but requires them to be drawn; ... To describe right lines and circles are problems, but not geometrical problems. The solution of these problems is required from mechanics; and by geometry the use of them, when so solved, is shown; and it is the glory of geometry that from those few principles, brought from without, it is able to produce so many things. Therefore geometry is founded in mechanical practice, and is nothing but that part of universal mechanics which accurately proposes and demonstrates the art of measuring.

The development of formal axiomatics and the new concept of mathematical theories, represent the second development just mentioned. Theories became forms, that is, they became recognized as realities in their own right, because of the wide range and great diversity of intended applications (be they mathematical or non-mathematical). The transformation of the very notion of axiomatics from Euclid to Hilbert testifies it. And David Hilbert made a remark "which contains the axiomatic standpoint in a nutshell: It must be possible to replace in all geometric statements the words point, line, plane by table, chair, mug" (Reid, 1970, p. 264).

Hilbert's remark is usually interpreted as expressing the tendency towards a de-ontologization of modern axiomatized mathematics. This is not so. Any formal theory has various intended applications or non-isomorphic models, and what the axioms describe are classes of objects rather than particular objects themselves. In this respect, mathematical axioms resemble natural laws. And like the latter, they have to be supplemented by an indication of the domain of objects to which they apply. And this is what Hilbert meant! (Hilbert, 1970, pp. 378ff).

Axiomatic theories are intentional theories: the axiomatic schemata define concepts, not objects and the concepts determine their referents. Anything that obeys Peano's axioms must be called natural numbers (Otte, 2003, p. 204). Set-theoretic mathematics, in contrast, is extensional: both aspects are complementary to each other. This comes out very clearly when we look at Peano's axiomatization of arithmetic. On two counts, Russell says, for example, Peano's approach

> fails to give an adequate basis for arithmetic. In the first place, it does not enable us to know whether there are any sets of terms verifying Peano's axioms. ... In the second place. ... we want our numbers to be such as can be used for counting common objects, and this requires that our numbers should have a *definite* meaning, not merely that they should have certain formal properties. (Russell, 1998, p. 10)

And further:

> If we start from Peano's undefined ideas and initial propositions, arithmetic and analysis are not concerned with definite logical objects called numbers, but with the terms of any progression. We may call the terms of *any* progression 0, 1, 2, 3,..., in which case 0, 1, 2, become "variables." To make them constants, we must choose some one definite progression; the natural one to choose is the progression of finite cardinal numbers as defined by Frege. (Russell 1998, p. 9; Russell, 1954, p. 4)

Russell's criticism was that the axiomatic characterization of number leads to a situation where, "every number-symbol becomes infinitely ambiguous" (Russell, 1998). Neither Peano nor Hilbert, according to Russell, are really capable of defining what *the* number One is. Russell seemed, however, not to

> have perceived clearly that Peano's procedure was a very general method of mathematics and that it was therefore in need of a set-theoretical foundation, if settheoretical thinking was to be—as claimed—an omnicomprehensive basis for all mathematics.
>
> On the other hand strictly formalistic mathematics, as it was developed by Hilbert's school, did not pay sufficient attention to all that burden of set-theoretic tools which were strictly connected with axiomatics and which can be summarized in the word "model." (Casari, 1974, p. 52)

So set-theoretical models and axiomatic systems jointly represent the essential complementarity of extensional and intentional aspects of mathematics. But this has so far hardly been recognized.

MATHEMATICS AS A LANGUAGE AND THE AXIOMATIC METHOD

Since the days of Plato's quarrels with the Sophists, mathematics and language has stood in certain opposition to each other. Language has two different sides, a representational and a communicative one, as we have seen already and Plato showed interest almost exclusively in the former. Privileging the cognitive aspects of language, represents not only a rejection of the confusing voices of myth, but suggests also that the philosophers should win over the rhetoricians, who try primarily to influence people and to drive their opinions in certain directions.

Still in the 16th and 17th centuries, mathematics and rhetoric were contrasted and the educational value of mathematics was considered negligible, while a close connection between mastery of rhetoric and good citizenship seemed indubitable. And when, as in the writings of Europe's first great political philosopher Thomas Hobbes, mathematics seemed of some relevance to the formation of the *vir civilis,* it was geometry and objective logic, rather than the dominant algebra and arithmetic, which were valued.

> Symbols are poor unhandsome, though necessary, scaffolds of demonstration.... Symbols, though they shorten the writing, yet they do not make the reader understand it sooner than if it were written in words. For the conception of the lines and figures (without which a man learneth nothing) must proceed from words either spoken or thought upon. (Hobbes in Molesworth, 1845/2013, pp. 248, 329)

Geometry's place in the picture of knowledge was originally bound up with its special truth status, due to the fact that the objects of geometrical reasoning were clear and distinct; while in analysis and algebra it was often unclear about what the symbols meant. In Euclid's *Elements,* Hobbes believed to have discovered a demonstrative science; one that could for the first time explain the true foundations of political science and justice (Skinner, 1996).

Until about 1800, mathematics consisted mainly in problem solving; the search for more powerful methods and new applications ruled the scene. Since then mathematics has become valued as a language not least because of the attempts to teach students how to use a coherent system of mathematics (Effros, 1998). The pedagogical principles underlying mathematics instruction, says Effros, "are quite similar to those used in language instruction" (p. 135). In educational contexts, problems play an auxiliary and instrumental role only. We do not, "include algebra in the high school curriculum in order to enable students to solve word problems" (p. 135).

But what is algebra? It is, on the one side, considered to be nothing but an arithmetical language, a view that Freudenthal had called anti-didactical; and it is, on the other side, a system of structures by the complementarity of operational and representational aspects of mathematical theories (Otte, 1994b). This becomes understandable as soon as we see the connection between algebra and modern axiomatics. The axiomatic method came about as soon as one realized that one was free to create all kinds of algebraic systems in which the variable stood for completely new objects. The works of Evariste Galois, Hermann Grassmann, A De Morgan, George Boole, and Benjamin Peirce, to name just a few, have provided a clear idea of this new algebraic spirit.

The axiomatic method has become a true expression of the complementarity of the descriptive or heuristics (extensional) and deductive (intentional) aspects of a theory, as soon as we understand that a theory is a pair consisting of a formal structure and a number of set-theoretical models or intended applications.

The following proof of the theorem of Pythagoras (Figure 8.1) provides a very simple illustration of this relative independence of structure and model.

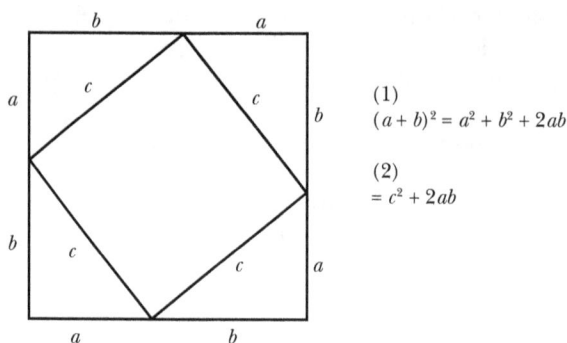

Figure 8.1

In the equation (1) the expression on the left side is simply calculated according to the laws of arithmetic; while in (2) one compares areas. Comparing (1) und (2) gives the desired result:

$$a^2 + b^2 = c^2$$

We have assumed here that the structure of algebra could be interpreted with respect to the number model, or equally well in terms of a calculation with geometrical segments. This independence of structure and model was Descartes' great discovery, which later became interpreted in a reductionist manner by the movement of arithmetization of mathematics.

What had dominated the view of mathematics as a language from the very beginning, however, was exactly the view of arithmetization (Klein, 1895). Arithmetization as a foundational science of mathematics does not mean that the actual subject matter of mathematics should be arithmetic. Rather numbers are no longer interpreted as objects, but are pure symbols and are means of objectifying mathematical thought. And arithmetic as a language is here to be taken exclusively in its communicative aspects, rather than as a means of objective representation. This identification of mathematical knowledge with its language, has very negative effects as it transforms mathematics education into a collection of rhetoric lessons. Discursive practices do not provide an object that really exists for us (Otte, 1994a, pp. 311ff). School mathematics is reducing algebra to arithmetic, and set-theory becomes a series of lessons in rhetoric and politics.

A PROBLEM IN THE PHILOSOPHY OF MATHEMATICS

There has always been a close connection between philosophical orientations and scientific theories, because it is philosophy that is responsible for our convictions on basic ontological questions. Many consider the main problem in the philosophy of mathematics to be the question: In what sense do mathematical objects exist? Reuben Hersh for example, is among those (Hersh, 1997). Until about 1800, he says, western philosophy believed that there were two kinds of things in the world: mental and physical. Hersh thinks that modern mathematics shows the inadequacy of this belief; and he proposes instead to consider mathematical objects as social entities and to acknowledge that mathematics is essential a social reality. Hersh wants to avoid the alternative of mentalism versus empiricism. Social entities, he says, "are neither mental nor physical," but they have "mental and physical aspects" (Hersh, 1997). A prominent author and mathematician, he seems right in some sense.

Hence comes the next question: What is human society, really? Since the Industrial Revolution, the answers to this question have commonly been framed in terms of two alternative schemas of comprehension: the paradigm of language, and the paradigm of production (Markus, 1986), such that we encounter the above complementarity reinforced. Both paradigms share some core ideas:

> first of all that of intersubjective constitution of subjectivity through those processes of objectivation (and reification) in which the individuals are involved and enmeshed in their daily commerce with each other and with the world of their common life. Nevertheless in some—and crucial—respects the two paradigms stand opposed to each other. Whether the relations of intersubjectivity are modeled on those pertaining to linguistic communication or on the interaction of individuals in the reproduction of their material livelihood—this choice has not only theoretical consequences. (Markus, 1986, p. xii)

And Markus devotes the first part of his book to a comprehensive critique of the so-called paradigm of language, as it has been represented by recent analytical philosophy. He points out that this paradigm leads to falling back into some kind of utter fatalism. The paradigm of production, on the other hand, is able to lend a more balanced viewpoint on the tension between the "human condition" and the "human striving" for a better society.

It seems a commonplace that modes of production profoundly influence all ways of social and cultural life. But differences in the mode of communication are often as important as differences in the mode of production; they do involve development in the creation and storing of human knowledge.

Depending on our sociological conceptions, we might connect mathematics either with communicative or with technological practice. We end up with mathematics as a language, or with a belief in the indispensability of mathematics to the empirical sciences and technologies.

NOTES

1. See the work of Claude Lévi-Strauss with respect to this dialectic, especially Lévi-Strauss, 1969, p. 109.
2. The references given (219, c-d, etc) point to the standard numbering of sections in Plato; see for example Cooper (1997).
3. See Plato (1926); for an online version see: Cratylus by Plato. Translated by Benjamin Jowett. Retrieved from: http://www.ancienttexts.org/library/greek/plato/cratylus.html
4. See for example http://en.wikipedia.org/wiki/The_Assayer

REFERENCES

Alexander, A. R. (2001). Exploration mathematics: The rhetoric of discovery and the rise of infinitesimal methods. *Configurations, 9*(1), 1–36.

Belhoste, B. (1981). *Augustin-Luis Cauchy.* New York, NY: Springer.

Bellarmino, R. (1615). Letter to Father Foscarini of April 1615. [Correspondence]. Retrieved from http://law2.umkc.edu/faculty/projects/ftrials/galileo/letterbellarmine.html

Bochner, S. (1966). *The role of mathematics in the rise of science.* Princeton, NJ: Princeton University Press.

Burtt, E. (2003). *The metaphysical foundations of modern science.* New York, NY: Dover.

Casari, E. (1974). Axiomatical and set-theoretical thinking. *Synthese, 27,* 49–61.

Cavaillès, J. (1976). *Sur la logique et la théorie de la science* [On the logic and theory of science]. Paris, France: Vrin.

Cooper, J. M. (Ed.). (1997). *Plato. Complete works.* Indianapolis, IN: Hacket Publishing Company, Inc.

De Condillac, E. B. (1746/2001). *Essay on the origin of human knowledge.* [Trans. & Ed. by H. Aarsleff]. Cambridge, England: Cambridge University Press.

Descartes, R. (orig. 1637). *Discourse on the method of rightly conducting one's reason, and of seeking truth in the sciences.* Retrieved from http://www.gutenberg.org/files/59/59-h/59-h.htm

Effros, E. G. (1998). Mathematics as language. In H. G. Dales & G. Oliveri (Eds.), *Truth in mathematics* (pp. 131–146). New York, NY: Oxford University Press.

Enriques, F. (1927). *Zur Geschichte der Logik* [On the history of logic]. Leipzig, Germany: Teubner.

Forster, G. (1785/1983). *Works in two volumes.* Berlin, Germany: Aufbau-Verlag.

Foucault, M. (1973). *The order of things.* New York, NY: Vintage.

Hahn, R. (1988). *Kant's Newtonian revolution in philosophy.* Carbondale, IL: Southern Illinois University Press.

Hankins, T. L. (1985). *Science and the Enlightenment.* Cambridge, England: Cambridge University Press.

Hersh, R. (1997). *What is mathematics, really?* New York, NY: Oxford University Press.

Hilbert, D. (1970). *Gesammelte Abhandlungen, Bd. III* [Collected papers, Vol. 3]. Heidelberg, Germany: Springer Verlag.

Israel, G. (1981). Rigor and axiomatics in modern mathematics. *Fund. Scientiae, 2*, 205–219.

Kant, I. (1787/1990). *Kritik der reinen Vernunft - 2. Auflage* [Critique of pure reason]. Hamburg: Felix Meiner Verlag.

Klein, F. (1895). Über Arithmetisierung der Mathematik [On the arithmetization of mathematics]. *Nachrichten der Königlichen Gesellschaft der Wissenschaften zu Göttingen, Geschäftliche Mitteilungen*, Heft 2.

Lévi-Strauss, C. (1961). *Tristes tropiques*. (J. Russell, Trans.). New York, NY: Criterion Books.

Lévi-Strauss, C. (1969). *The elementary structures of kinship*. London, England: Eyre & Spottiswoode.

Markus, G. (1986). *Language and production: A critique of the paradigms*. Dordrecht, the Netherlands: Reidel.

McMullin, E. (1998). Galileo on science and scripture. In P. Machaumer (Ed.), *Cambridge companion to Galileo* (pp. 271–347). Cambridge, England: Cambridge University Press.

Molesworth, W. (Ed.). (1845/2013). *The English works of Thomas Hobbes of Malmesbury. Vol. VII*. London, England: Longman, Brown, Green, and Longmans. Retrieved from www.forgottenbooks.org

Mueller, I. (1969). Euclid's elements and the axiomatic method. *British Journal for the Philosophy of Science, 20*, 289–309.

Nahin, P. (1998). *An imaginary tale. The story of $\sqrt{-1}$*. Princeton, NJ: Princeton University Press.

Otte, M. (1989). The ideas of Hermann Grassmann in the context of the mathematical and philosophical tradition since Leibniz. *Historia Mathematica, 16*, 1–35.

Otte, M. (1991). Style as a historical category. *Science in Context, 4*, 233–264.

Otte, M. (1993). Towards a social theory of mathematical knowledge. In C. Keitel & K. Ruthven (Eds.), *Learning from computers* (pp. 280–306). Berlin, Germany: Springer.

Otte, M. (1994a). Mathematical knowledge and the problem of proof. *Educational Studies in Mathematics, 26*, 299–321.

Otte, M. (1994b). *Das Formale, das Soziale und das Subjektive* [The format, the social and the subjective.] Frankfurt, Germany: Suhrkamp.

Otte, M. (2003). Complementarity, sets, and numbers. *Educational Studies in Mathematics, 53*, 203–228.

Otte, M., & Zawadowski, W. (1985). Creativity. *Educational Studies in Mathematics, 16*, 95–97.

Paulos, J. A. (1998). *Once upon a number: The hidden mathematical logic of stories*. New York, NY: Basic Books.

Peirce, C. S. (1958–1966). *Collected papers*. Vols. 1–6, C. Hartshorne & P. Weiss (Eds.); Vols. 7–8, A. W. Burks (Ed.). Cambridge, MA: Belknap.

Plato (1926). Cratylus. Parmenides. Greater Hippisas. Lesser Hippias. Transl. Harold North Fowler. Plato Volume IV. Loeb Classical Library 167. Cambridge, MA: Harvard University Press.

Reichenbach, H. (1951). *The rise of scientific philosophy*. Berkeley, CA: University of California Press.

Reid, C. (1970). *Hilbert*. Heidelberg, Germany: Springer.

Rousseau, J. J. (1754) *Discourse upon the origin and the foundation of the inequality among mankind*, (Kindle ed.).

Russell, B. (1954). *The analysis of matter.* London, England: Allan & Unwin.

Russell, B. (1998). *Introduction to mathematical philosophy.* London, England: Routledge.

Skinner, Q. (1996). *Reason and rhetoric in the philosophy of Hobbes.* Cambridge, England: Cambridge University Press.

Weyl, H. (1966). *Philosophie der mathematik und naturwissenschafte* [Philosophy of mathematics and natural science]. München, Germany: Oldenburg.

Wulf, A. (2012). *Chasing Venus: The race to measure the heavens.* London, England: Heinemann.

Michael Otte
Universität Bielefeld, Germany, and
Universidade Anhanguera de São Paulo (UNIAN), Brazil

CHAPTER 9

(ETHNO)MATHEMATICS AS DISCOURSE

Ole Skovsmose

INTRODUCTION

Ethnomathematics can refer to different practices that include mathematics.[1] It can also refer to a research approach, and one can talk about the ethnomathematical research program.[2] In this chapter, I will concentrate on ethnomathematics as referring to a variety of practices. Although including some remarks about ethnomathematics as a research programme.

Several ethnomathematical studies indicate a duality between ethnomathematics and academic mathematics. On the one hand, academic mathematics has been described as a dominant regime of truths, which defines standards according to which other ways of thinking about numbers, magnitudes, forms, space, and time, become inadequate, if not simply wrong. Ethnomathematics, on the other hand, is described as integrated in many different everyday practices and as being made part of lived-through cultured values. While academic mathematics becomes characterized mainly through negative terms, ethnomathematics is presented in mainly positive.

One purpose of this paper will be to challenge any such duality. I will try to show that we do not need any good-bad dichotomy. However, the

Refractions of Mathematics Education, pages 155–172

purpose of the paper will be broader. I will: (1) reflect on the use of the notions of mathematics and ethnomathematics; (2) characterize a discursive interpretation of language; (3) formulate the thesis that mathematics is discourse; (4) substantiate this thesis by presenting some mathematics-based discursive acts; (5) reformulate the argument by considering ethnomathematics-based discursive acts; and (6) present a critical perspective on both mathematics and ethnomathematics.

MATHEMATICS AND ETHNOMATHEMATICS

According to the classic referential interpretation, the meaning of a notion is the entity to which it refers. This interpretation has been discussed, elaborated on, and revised throughout the history of philosophy. Augustine proposed such a general referential interpretation of meaning. And Gottlob Frege, to make a huge jump forward in history, elaborated a referential theory with particular reference to mathematics (1967, 1978). Frege's work established a modern version of Platonism, and in many cases a referential theory of meaning that incorporated features of Platonism.

In *Philosophical Investigations*, Ludwig Wittgenstein (1953/1958) challenged any referential interpretation of meaning. He made explicit reference to Augustine, and he was completely familiar with Frege's work. So Wittgenstein knew what he was up against. Contrary to Frege, he found that it did not make sense to clarify the meaning of a number concept, say 2, by identifying what 2 in reality referred to. Concepts did not have any real reference. Thus it did not make sense to follow any Platonic approach by clarifying not only mathematical notions but also notions (like beauty, justice, or truth) by identifying their ideal references. Sure, it was difficult to address questions like: What is a number; what is mathematics, really; what is art; what is the good action; and what is knowledge. However, according to Platonism it was the task of philosophy to address such questions. And even if Platonism was not explicitly assumed, one found many attempts to address such questions, assuming that one in fact was looking for a particular answer.

Wittgenstein's point in *Philosophical Investigations* was that there is not anything particular deep in any questions of the form: What is X really? The search for the meaning of a notion was not a question of excavating its essence. Concepts had no essence. There was no ideal world of ideas, or references, or essences that was waiting to be discovered. The existence of any such Platonic "eternity" with respect to meaning was a myth.

According to Wittgenstein, we have to look for the meaning of a concept in a quite different way. The meaning has to be located in real-life practices. The meaning of a concept is social and can be associated to the *use of the concept*. Wittgenstein emphasized this point by introducing the notion of "language game." A concept can be made part of different language games; and by exploring such game, one might grasp the meaning of a concept. Some games might be rather similar and demonstrate family resemblances. Language games were cultural products. They could be quite stable during some periods, and made changes during other periods. Language games could proliferate; they could combine; they were dynamic; and so were the meanings of concepts.

I follow Wittgenstein in this interpretation of meaning. This also applies to a notion like mathematics. This is not a notion with any proper or principal meaning. Mathematics can be part of many different language games. It can refer to different practices, and we can talk about academic mathematics, pure mathematics, school mathematics, applied mathematics, engineering mathematics, insurance mathematics, mathematics of finance, or others. There is no end to the different possible uses of mathematics. And just as there is a plurality of uses, there is a plurality of meanings for mathematics.

In a similar way, I use ethnomathematics broadly and freely as referring to many different practices. It can refer to shoemakers' mathematics, tailors' mathematics, brick builders' mathematics, sugar-cane farmers' mathematics, etc. Following Ubiratan D'Ambrosio's suggestion of using a broad notion of ethnomathematics, one could also refer to engineering mathematics, academic mathematics, insurance mathematics, and others as examples of ethnomathematics.

In fact, I do not find any particular need for distinguishing between mathematics and ethnomathematics. In this chapter, I am going to substantiate this claim by showing that a discursive interpretation of language, can be applied to both mathematics and ethnomathematics. However, let us proceed step by step.

A DISCURSIVE INTERPRETATION OF LANGUAGE

One can meet very different interpretations of language. A most general *representational interpretation* is formed through Wittgenstein's picture theory of language, as suggested in *Tractatus* (Wittgenstein, 1921/1992). The picture theory includes the general claim that language provides a more or less accurate description of reality. Furthermore, *Tractatus* included the idea that some languages do the picturing better than other languages. In fact Wittgenstein talked about *the* language, and not about languages in plural. According to Wittgenstein, *the* language that was adequate for picturing reality, was the formal language of mathematics. This language was presented in *Principia Mathematica* by Alfred N. Whitehead and Bertrand Russell (1910–1913), but prepared in detail by Frege.

A radically different interpretation of language can be referred to as a *discursive interpretation* of language. Such an interpretation can be seen as a further development of Wittgenstein's notion of language game. Thus, while *Tractatus* included one interpretation of language, *Philosophical Investigations* opened up a very different one. Thus Wittgenstein was the main figure in two competing interpretations of language.

I will characterize a discursive interpretation of language in terms of three features: the first includes strict anti Platonism. We have already indicated what this could mean with respect to interpretations of concepts: there would be no real reference that established their proper meanings. Instead the meaning of a concept is defined through real-life practices: the use of the concept. More generally, the anti-Platonism that characterizes a discursive interpretation of language, includes a negation of any attempts to think of language as providing a representation of reality. A discursive interpretation does not operate with any sharp distinction between language and reality. Instead it opens different forms of language-reality transitions: reality is discursively constructed, while language is formed through real-life practices. So the first feature of a discursive interpretation of language is the recognition of the many forms of language-reality transitions.

The second feature of a discursive interpretation of language, is the recognition of the action part of language. One finds such a recognition in Wittgenstein's notion of language games, as playing means action. A more explicit formulation of the action part of language is provided by John L. Austin (1962, 1970). He pointed out that a statement had a locutionary content, an illocutionary force, and a perlocutionary effect. As an example, we can imagine that somebody warns me: Do not to make business with that company. The locutionary content refers to the content of the warning, that having to do with not making business. The illocutionary force refers to the power of the statements: there is a fact made into a warning. Finally, the statement has a perlocutionary effect, which refers to the effects that the warning might have: thus I might feel surprised by the warning. Austin's point was that any statement included these three features, including promises, proposals, excuses, invitations, demands, critique, and others. We *do things* through language. The speech act theory was further developed by John Searle (1969), in his book *Speech Acts*. After which the speech act theory proliferated, and became generally acknowledged: language includes actions.

The third feature of a discursive interpretation of language, is the recognition of the political dimension of language. As an illustration of what this could mean one can listen to the following remark made by Slavoj Žižek:

> Language simplifies the designating thing, reducing it to a single feature.
> It dismembers the thing, destroying its organic unity, treating its parts and

properties as autonomous. It inserts the thing into a field of meaning which is ultimately external to it. (Žižek, 2008, p. 61)

Certainly Žižek recognizes the action part of language. However, his formulations are much more radical than Austin's peaceful and innocent examples that have been elaborated on further in analytical philosophy. Through this philosophy, we come to learn about invitations, warnings, and promises. But not about exclusion, stigmatization, and certainly not anything about the symbolic violence that is Žižek's topic in the book *Violence*. Symbolic violence represents a big part of the political dimension of language. To recognize this, means to recognize that interests, subjectivity, and priorities of any kind may be acted out through language; not only through the explicit statements, but also through the world view that might be engraved in language.

When one talks about particular interests, subjectivity, and priorities, then one might assume that it becomes possible to identify general interest, objectivity, and neutrality. However, the discursive interpretation of language does not make any such assumption. Recognizing the political dimension of language also means recognizing that there might not exist within language any semantic fixed points; a reference to which one can judge degrees of interest, subjectivity, and priorities.

MATHEMATICS AS DISCOURSE

My proposal is to think of mathematics as language, and to establish a discursive interpretation of this language. I suggest that we think of mathematics as discourse.[3] This means that we do not think of mathematics, nor of mathematical modeling, as a straightforward device for representations.[4] Mathematics is not any sublime means for picturing reality, as has been suggested in much literature about mathematical modeling. I suggest instead that we recognize the possibility of a broad variety of mathematics-reality transitions; that we think of mathematics as including a dimension of action; and that we acknowledge that mathematics includes a political dimension. Through mathematics one engages in reality: the *homo faber* is operating with and through mathematics. Through mathematics, one can impose certain interests, a particular perspective, a particular world-view. One might as well be able to associate symbolic power, as well as symbolic violence, with mathematics.

In the following, I will try to substantiate that the features of the discursive interpretation of language, apply to mathematics. However, before we get into this, let me make a preliminary observation. As indicated in the beginning, the very notion of mathematics cannot be assumed to have any proper and universal meaning. This, however, does not prevent me, and should not prevent us, from using the notion; and also using it for making general claims like, mathematics can be seen as discourse. However, making such a general statement, does not imply that one has tried to say something about the essence of mathematics. Making a general statement, means that one has tried to say something that makes sense for several instantiations of mathematics. General statements are tentative, they are preliminary, they are suggestions; they can be seen as guesses, and also as bold guesses. Having this in mind, let us proceed.

MATHEMATICS-BASED DISCURSIVE ACTS

Mathematics operates within a range of social, political, and economic structures in society. Thus Eva Jablonka and Uwe Gellert (2007, p. 20) emphasize that mathematics-based decisions operate on many levels:

> On the level of national policy, decisions about the distributions of state salaries, pensions, and social benefits rely on mathematical extrapolations of demographical and economic data...On the level of interpersonal relations, mathematics-based communication technologies have already changed the habits and styles of private conversations.

One could elaborate further on this comment and refer to: administrative procedures, organisational schemes, health care programmes, and military operations.

I will try to substantiate this general claim by elaborating further on the discursive interpretation of mathematics. I will identify some mathematics-based discursive acts by addressing the formation of: (1) possibilities, (2) rationality, (3) structures and artifacts, (4) authority, and (5) ways of overlooking.[5]

Formation of Possibilities

A characteristic feature of modern technology is the conception of new possibilities for making constructions of any kind of artifact: buildings, bridges, ships, aircrafts, cell phones, and others. One can also think of the formation of, for instance, production processes, management procedures, information processing, and schemes for surveillance. For any such construction or formation, it is important to provide a blueprint of what to make. And mathematics is essential for being able to provide this. In all forms of technology, mathematics-based blueprints can open new possibilities. The hectic development taking place in computing and information technology is an example of this.

Naturally one might consider the extent mathematics also limits the space of technological possibilities, by eliminating what could not be formulated in this language. In this sense, mathematics also plays a defining role in technological development. The very formation of technological possibilities is a principle example of a mathematics-reality transition. This also is a direct example of a mathematics-power interaction.

Formation of Rationality

Often in political discussions it has been argued that certain economic decisions need to be taken, in order to perhaps maintain economic stability. Most often, the output from a mathematical model has structured such arguments. Models for economic forecasting are applied by governments, political parties, research institutions, banks, and companies.

A model of this kind is loaded with information based on statistics. It is also designed through a huge number of equations, which represent mathematical insight, economic priorities, political perspectives, and particular market interests. Altogether the components of the model establish an economic universe, which is used for economic forecasting. Certain values become assigned to certain parameters; hypothetical implications become identified by running the model. A broad experimentation with different possible inputs, results in a reading of economic alternatives and their implications. On this basis, one might claim that some necessary economic initiatives have been identified. This "necessity", however, is a constructed necessity; it has been fabricated by means of the model. The model itself is not any economic reality, but it plays the role of reality; it establishes a symbolic reality as a parallel economic universe. The "necessary" actions are identified within this parallel universe, but applied in real life. In this sense, we have a model-based formation of an economic rationality.

A similar formation is found in management and engineering. One finds this formation when new forms of production are implemented; when possibilities for outsourcing are discussed; and when promotion campaigns are launched. One finds a mathematical formation of rationalities in medicine, as well as in warfare. Such formations can produce actions of profound political significance.

Formation of Artifacts and Structures

If we look at any industrial produced artifact—shoes, cars, computers, plastic bags, etc.—mathematics has been crucial for their design and production.

Let us just consider the industrially produced plastic bag that I use when going to buy fruits and vegetables. It is a colourful bag, woven of plastic stripes. It is very strong, and I can put in as much as I am able to carry. It is an industrially produced bag, and I try to imagine the machinery that produces such bags. I remember a section in the Deutsches Museum in Munich that showed the history of weaving. Many tools were used before the first spinning machine was constructed in 1764. In fact it was surprising to see the great homogeneity of tools that were applied century after century in this production; and also to see that the first weaving machine not only represented a discontinuity in this development, by being a machine, but also a big homogeneity that applied the usual tools. The machines, however, were rapidly developed; machine power was added, and the weaving machines took on tremendous dimensions. Later computational techniques were added, and the automation became tremendously sophisticated. My plastic bag was produced by some machinery of this kind; a machinery that represents a complex mathematics-based development.

However, not only artifacts or the very structures of production, have been formed through mathematics. Let me refer to one specific feature of production, namely robotting.[6] Some initial steps in the configuration of workers as robots were presented by Frederick W. Taylor in *The Principles of Scientific Management*. Here Taylor (1911/2006) describes how the worker named Schmidt was subjected to the principles of scientific management. Schmidt was loading pig iron onto a railway car, and it was observed what an average worker was loading per day. Taylor and his staff had, however, analyzed the work process in all its parts, and came to the conclusion that a new way of organizing the whole process would more than triple the loading efficiency. A new working algorithm, which prescribed what do to and when to do it, was presented to Schmidt. In this way, he was configured as a robot.

Taking a look at any industrial production today, one can find human robots and proper robots united in the complex processes of production. The distinction between what is completed by human beings and what is done by machines, is not easy to maintain. In fact, in terms of efficiency, it is irrelevant to make any such distinction, as it is the combined machinery that is optimized. The whole configuration of production could only take place through an extensive use of mathematical modelling, including mathematics-based cost-benefit analyses. In this way, mathematics defines the whole formation of structures of production. This also applies to the production of my colourful bag.

Formation of Authority

Mathematics-based discursive acts include a reconfiguration of authority. Claims expressed in numbers more easily turn powerful. This is an example of the symbolic power exercised by politicians, governments, and institutions that have access to mathematical models.

However, authority is not only formed through the power of argument, it is also established through the power of control. As a particular example, one can consider the technology of surveillance.[7] Surveillance means observing in order to control what is taking place. However, efficient surveillance also means remembering or registering what has been observed.

Surveillance is crucial for making a population governable. One can register names of people, their actions, and (in particular) their criminal acts. One can register people's income, debt, fortune, all kind of information important for determining and controlling tax payments. One can register people's health conditions, and try to make medical treatment more effective. One can register opinions and political priorities, and then try to make political campaigns more powerful. One can register students' performances, and on that basis try to establish efficiency with respect to the distribution of students in streams of further education; and in the end form a more productive future work force. One can register workers' performances, and try to optimize the production process.

All such forms of registering are mathematics based. They are crucial to making a population governable; it is part of the formation of authority.

Formation of Overlooking

Through mathematics one can see many things. At the same time mathematics represents a system of ignoring and overlooking; for example, mathematics-based scientific management. This management imposes a mathematical perspective on the process of production. One comes to see production as a mechanical process that one can try to optimize; workers are configured as elements in this mechanical process.

This configuration also means that many features of being a worker become ignored: that the worker has a family, has responsibilities, has personal interests and priorities. These are not part of the mathematics-based representation of the worker as an element in the production process; in general, mathematics does not represent human beings as being human. Instead mathematics facilitates a mechanical perspective on things, be that nature, economic issues, socio-political affairs, or human beings. In all such cases, mathematics operates as a means for both seeing and overlooking.

ETHNOMATHEMATICS-BASED DISCURSIVE ACTS

The previous section dealt with mathematics-based discursive acts in terms of formation of possibilities, rationality, artifacts and structures, authority, and overlooking. I tried to explore these acts with reference to examples of what broadly could be referred to as engineering mathematics. In this way I have tried to illustrate that mathematics-based acts transgress any language-reality dichotomy; that they forms actions, and that they include a political dimension. In other words, I have tried to show that a discursive interpretation of mathematics makes sense.

But what about ethnomathematics? I will try to illustrate that it makes good sense to talk about ethnomathematics-based discursive acts, and that they form possibilities, rationality, structures and artifacts, authority, and ways of overlooking.

Formation of Possibilities

As an example of formation of possibilities, one can think of travelling and navigation. If we consider the period of the so-called big discoveries, more or less reliable maps were constructed; and before any mapping could be done, one tried to identify a route or a direction. One could go west, but what would that mean in a boat in the middle of the ocean? Many tools for navigation were developed; not only the compass, but also the sextant, and a wide range of calculation devices.

We can go further back in time, and think about the Vikings who reached Iceland, then Greenland, and finally America. They navigated without a compass, but they certainly used a range of techniques and tools for reading the stars. Thus the firmament of stars was a rather stable, although rotating, dome above the all too capricious ocean. We can also consider the crossing of the ocean by the Polynesians. They populated the thousands of islands in the Pacific, and demonstrated a most remarkable capacity for navigating.

The mathematics involved in navigating, provides several examples of ethnomathematics; and one can think of navigation as an example of ethnomathematics-based discursive acts.[8] Navigation does not include any sharp distinction between mathematics and reality. Quite the contrary, navigation represents a complex mixture of interpreting and doing; it also represents a mathematics-based formation of possibilities. Similar remarks could apply as well to any form of navigation today, including the most advanced systems operating in space missions, as well as in GPS used when tourists try to find their way around.

Formation of Rationalities

Formation of rationalities has to do with reasons and conditions for doing something, and for doing something in a particular way. Let me remark about procedures for dividing. Making a division is a social process. It takes place in families, in communities,and in a larger context. It is a process that is transformed into extreme complexity in tax systems and in public welfare systems. Naturally, it is an open question what is considered a fair division. One can think of the mathematical algorithm for division as a formal suggestion, although very simplistic, for what fairness could mean.

Let me refer to an example related to me by a researcher in ethnomathematics: Members of an indigenous group of Indians were presented for the following question. Three men were out fishing together. They caught in total 36 fishes. How many would each get? The Indians looked at each other and asked: What kind of fish was it?

To mathematics teachers, this answer might sound funny. But it could instead be seen as demonstrating the complexity of the notion of fair division. This notion is far from always grasped by mathematical division. For instance, the division could concern very small fishes that had to be prepared all together; or big and tasty fishes that need to be distributed according to the size of the families. The mathematical algorithm of division provides one interpretation of what a fair division could mean; but it is just one interpretation that might overlook parameters that also needed to be considered.

One can see several ethnomathematical approaches as attempts to provide rationales for making fair divisions. As an example, one can think of the techniques for measuring lands that are developed for sugar-cane farming (Abreu, 1993). One aim of this measuring has been to estimate what could be a fair total payment; another aim could be to make a fair division among the workers. And certainly, either time one has raised the question about what a fair division could mean.

Formation of Artifacts and Structures

Many ethnomathematical studies have addressed the construction of artifacts.[9] As an example, one can think of basket weaving. Through a range of studies, Paulus Gerdes (2008, 2012) demonstrated the mathematical competences that were represented in weaving practices. In his book *Otthava*, he analyzed such practices in the Makhuwa culture, located in northeast Mozambique.[10] Beyond the weaving of baskets and handbags, Gerdes investigated the weaving of hats, fish traps, containers, trays, mats and several other artifacts. He tried to identify the kind of knowledge that was acted out through weaving practices; those concerning, for example, symmetries, spirals, and cylinders. Thus Gerdes provided a profound study of the different layers of tacit knowledge that formed such practices, and among such layers he found mathematics.

Such research approaches, however, have been the object of heated discussions. Does that have to do with any application of mathematics? It has been claimed, instead, that there is a projection of mathematical ideas into the weaving process; the weaving process includes a range of procedures and competences that form a complex amount of tacit knowledge. But it is only the outsider, the onlooker, the mathematician, who can identify the mathematics in this tacit knowledge. Thus it has been claimed that the ethnomathematics included in weaving, is an invention; it is an external reconstruction. That the ethnomathematics in such practices has been falsely added.

Somehow, I do not find that this controversy needs to be properly clarified, at least not for the moment. Let us again consider my plastic bag. Does it include mathematics? Well, enormous amounts of mathematics have been put in operation with the development of weaving machines. Mathematics has contributed to part of the fabrication of the machines; in the organization of the production process, and in the computational automation of it. In this sense my plastic bag, is rich with mathematics.

Who is aware of this mathematics? Well, I do not think of such things when I use the bags; this is no surprise. The operator involved in the production then? Well, they are operating the machines: they are observing that everything works properly, they are checking the computer screens. I would, however, be surprised if the operators believed that they were doing mathematics. What about the engineers who configured the machines for this particular production? They might have used mathematics explicitly. However, basically, the engineering mathematics involved in the industrial production of plastic bags does not emerge in this situation. It remains as the tacit knowledge inscribed in the machinery that is in operation. This observation has to do with process of demathematisation, which accompanies the mathematisation of society.[11]

Naturally, one might claim that engineering mathematics was explicit at least at some stage in the bag-weaving process, while the basket-weaving mathematics were imposed on the situation. However, I do not find the difference between basket-weaving mathematics and bag-weaving mathematics to be that crucial. In both cases, we have a mathematics that includes processes of production.

Formation of Authority

Authority can be exercised through procedures and regulations: think of the procedures for taxation. Identifying the amount tax to be paid by a person, is far from just a descriptive statement; it is an act, it is a political act, and it represents power and authority.

More generally, authority is exercised through claims about how to do things. For example, one can think about the definition of procedures for production. We can find examples of modern and industrialized forms of production, but one can also think of many others. Let me refer to an example presented by Aldo Parra Sánchez (2012), who for a period stayed with groups of Indians in the Andes in Colombia. In particular, he followed the planting of corn, which during centuries has been the principal food ingredient in this region.

The planting and cultivation of corn emerged in routines and rituals. For instance, corn was planted, not in long and parallel rows, but in a huge spiral covering the whole field. The straight lines that we see in industrialized farming are machine made; while the spiral is handmade. But there might be more to the spiral than just handcraft. This way of organizing the planting can be interpreted as an understanding of what it means to engage with nature; the spiral is an expression of tradition and culture, combined into a certain ethnomathematics.

One can see the organization of production, as initiated by scientific management, as an expression of a particular metaphysics having to do with exploitation. One can see the organization of the cultivation of corn also as an expression of metaphysics, although a quite different one. Different as they are, both cases can be read as examples of formation of authority. In the case of the corn, we have an authority rooted in tradition, culture, and metaphysical ideas. And the ethnomathematical expression of this authority has provided a pattern for how to do things. In the case of industrial production, we also find an authority; here it is expressed through standards and algorithms for production, steered by criteria for maximizing profit. It is an authority of exploitation.

Formation of Overlooking

Ethnomathematical studies have concentrated on showing what to see and grasp through ethnomathematical insight and techniques; and maybe first of all, what one can do.

There are not many ethnomathematical studies that concentrate on revealing what an ethnomathematical perspective on a certain phenomenon might ignore and overlook. In fact, now thinking about it, no such studies come to my mind. Anyway, I see ethnomathematics, as mathematics in general; it forms not only what one sees, but also what one cannot see.

This issue, however, needs to be explored much further. There are many issues that one could consider. House building: Are there possibilities for construction that tend to be ignored, due to the ethnomathematical outlook? Basket weaving: It has been carefully studied, but in what sense do weaving techniques condense ethnomathematical insight and competence. Are there some such insights that represent limitations and obstructions for the weaving? Health care: Are there, due to the ethnomathematical perspectives, possibilities that have been ignored with respect to the potential of natural medicines that are available? Sugar-cane farming: It has been carefully studied how areas are measured and salary for field work calculated. But are there elements of such calculations that may serve the workers' interest, or serve the employers' interests?

I find it is important to provide a more complete picture of techniques that are rooted in ethnomathematics; that the possibilities of ignoring and overlooking needs to be broadly explored.

ACTS CAN HAVE ALL KIND OF QUALITIES

I have tried to show that discursive acts can be associated with both mathematics and ethnomathematics. In both cases, we can find formations of possibilities, rationality, structures and artifacts, authority, and overlooking. Considering particular examples of such discursive acts, we have found that they include a transgression of any mathematics-reality distinctions, that they include actions, and that such actions could be political.[12]

Both mathematics and ethnomathematics operate through discursive acts, and such acts, like any other kind of acts, can have any qualities. They might be efficient, misguided, expensive, risky, authoritative, benevolent, suppressive, dubious, and more. This means that both mathematics and ethnomathematics need to have equally profound critical approaches. A general critique of mathematics has been formulated, for instance through the discussion of mathematics in action and the formatting power of mathematics. However, it is equally important to provide a careful critique of ethnomathematics in action, and the formatting power of ethnomathematics. As a consequence, I do not think it makes sense to elaborate on a duality between, say, ethnomathematics and academic mathematics, as referred to in the introduction. And certainly it does not make sense to try to associate a range of positive qualities to ethnomathematics, leaving dubious qualities for academic mathematics.

Let me now return to the very distinction between mathematics and ethnomathematics. I find it makes sense to provide a discursive interpretation of ethnomathematics as well as of mathematics in general. As a consequence, I do not see any particular reason for maintaining any distinction between mathematics and ethnomathematics. In fact I prefer to talk about mathematics. Keeping in mind my initial observation that mathematics is an open concept. Mathematics is without any Platonic kernel. We have a multitude of mathematics, and I can talk about the mathematics of making plastic bags as well as the mathematics of basket weaving. So I do not want to make any distinction between practices that include mathematics and practices that include ethnomathematics.[13]

Let me conclude with a remark about ethnomathematics as a research program. One can consider ethnomathematics as a way of looking at mathematics in any of its many instantiations (Fernández, 2004). It is a way of addressing the social dimensions of mathematics whatever kind of mathematics we have to do it with. Being so, I find the ethnomathematical research program to be crucial. It might open up a critical perspective on mathematics, for any kind of mathematical practices.

ACKNOWLEDGEMENTS

This article has been published in *BOLEMA*, Volume 29, Number 51, 2015, and I thank the editors of *BOLEMA* for kind permission to publish it in the present *Festschrift.* I want to thank Denival Biotto Filho, Renato Marcone, Raquel Milani, Aldo Parra Sánchez and Miriam Godoy Penteado for making suggestions and critical comments to the manuscript.

NOTES

1. This is, for instance, the use found in Ascher (1991).
2. For a presentation of an ethnomathematical research program, see D'Ambrosio (1992, 2006, 2008).
3. This proposal does not exclude, however, that one can think of mathematics as being other things as well.
4. This issue I have discussed more carefully in Skovsmose (2012b).
5. For other formulations of such acts see, for instance, Skovsmose (2009, 2010).
6. For a discussion of robotting, see Skovsmose (2012a) where robotting is defined as "a symbolic power exercised when a mathematical discourse grasps only the 'mechanisms' of work processes" (p. 128).
7. For a more detailed discussion of this example, see Skovsmose (2012a).
8. For an overview of Pacific ethnomathematics, see Goetzfridt (2007).
9. See, for instance, Costa, Catarino and Nacimento (2008a, 2008b); Giongo (2004); and Sousa, Palhares and Sarmento (2008).
10. See also Vieira, Palhares and Sarmento (2008).
11. For a discussion of matematisation and demathematisation see Gellert and Jablonka (2009), and Jablonka and Gellert (2007).
12. Many studies explicitly address the political dimension of ethnomathematics. See, for instance, Knijnik (1996, 2008) and Knijnik, Wanderer and Oliveira (2004).
13. Finally I got to this conclusion. However, I am far from the first getting. Thus the title of Costa, Catarino and Nacimento (2008a, 2008b) are "Tanoeiros em

Trás-os-Montes e Alto Douro: Saberes (etno)matemáticos" and "Latoeiros em Trás-os-Montes e Alto Douro: Saberes (etno)matemáticos". Like in my title they talk about (ethno)mathematics.

REFERENCES

Abreu, G. (1993). The relationship between home and school mathematics in a farming community in rural Brazil. (Doctoral dissertation). Cambridge, England: Cambridge University.

Ascher, M. (1991). *Ethnomathematics: A multicultural view of mathematical ideas.* Washington, DC: Chapman and Hall/CRC.

Austin, J. L. (1962). *How to do things with words.* Oxford, England: Oxford University Press.

Austin, J. L. (1970). *Philosophical papers,* (2nd ed.). J. O. Urmson & G. J. Warnock (Eds.). Oxford, England: Oxford University Press.

Costa, C., Catarino, P. A., & Nacimento, M. M. (2008a). Tanoeiros em Trás-os-Montes e Alto Douro: Saberes (etno)matemáticos [Coopers in Trás-os-Montes and Alto Douro: (Ethno)mathematical knowledge]. In P. Palhares (Ed.), *Etnomatemática: Um olhar sobre a diversidade cultural e a aprendizagem matemática* (pp. 193–233). Ribeirão, Portugal: Edições Húmus.

Costa, C., Catarino, P. A., & Nacimento, M. M (2008b). Latoeiros em Trás-os-Montes e Alto Douro: Saberes (etno)matemáticos [Tinsmiths in Trás-os-Montes and Alto Douro: (Ethno)mathematical knowledge]. In P. Palhares (Ed.), *Etnomatemática: Um olhar sobre a diversidade cultural e a aprendizagem matemática* (pp. 235–264). Ribeirão, Portugal: Edições Húmus.

D'Ambrosio, U. (1992). Ethnomathematics: A research program on the history and philosophy of mathematics with pedagogical implications. *Notices of the American Mathematics Society, 39,* 1183–1185.

D'Ambrosio, U. (2006). *Ethnomathematics: Link between traditions and modernity.* Rotterdam, the Netherlands: Sense.

D'Ambrosio, U. (2008). Globalização, educação multicultural e o programa Etnomatemática [Globalization, multicultural education, and the program of ethnomathematics]. In P. Palhares (Ed.), *Etnomatemática: Um olhar sobre a diversidade cultural e a aprendizagem matemática* (pp. 23–46). Ribeirão, Portugal: Edições Húmus.

Fernández, E.L. (2004). As matemáticas da tribo europeia: Um estúdio de caso [The mathematics of the European tribe: A case study]. In G. Knijnik, F. Wanderer, & C. J. Oliveira (Eds.), *Etnomatemática, currículo a formação de professores* (pp. 124–138). Santa Cruz do sul, Brazil: EDUNISC.

Frege, G. (1879/1967). Begriffsschrift, a formal language, modeled upon that of arithmetic, for pure thought. In J. V. Heijenoort (Ed.), *From Frege to Gödel: A source book in mathematical logic, 1879–1931* (pp. 1–82). Cambridge, MA: Harvard University Press.

Frege, G. (1978). *The foundations of arithmetic/Die Grundlagen der Arithmetik.* (Trans. by J. L. Austin). Oxford, England: Blackwell.

Gellert, U., & Jablonka, E. (2009). The demathematising effect of technology. In P. Ernest, B. Greer, & B. Sriraman (Eds.), *Critical issues in mathematics education* (pp. 19–24). Charlotte, NC: Information Age.

Gerdes, P. (2008). Explorando poliedros do Nordeste de Moçambique [Exploring polyhedrons in the Northeast of Mozambique]. In P. Palharas (Ed.), *Etno-matemática: Um olhar sobre a diversidade cultural e a aprendizagem matemática* (pp. 317–359). Ribeirão, Portugal: Edições Húmus.

Gerdes, P. (2012). *Otthava: Making baskets and doing geometry in the Makhuwa culture in the Northeast of Mozambique.* Morrisville, NC: Lulu.com.

Giongo, L.M. (2004). Etnomatemática e práticas de produção de calcados [Eth-nomathematics and practices of shoe production]. In G. Knijnik, F. Wande-rer, & C. J. Oliveira (Eds.), *Etnomatemática, currículo a formação de professores* (pp. 203–218). Santa Cruz do sul, Brazil: EDUNISC.

Goetzfridt, N. J. (2007). *Pacific ethnomathematics: A bibliographic study.* Honolulu, HA: University of Hawaii Press.

Jablonka, E., & Gellert, U. (2007). Mathematisation—Demathematisation. In U. Gellert & E. Jablonka (Eds.), *Mathematization and de-mathematization: Social, philosophical and educational ramifications* (pp. 1–18). Rotterdam, the Nether-lands: Sense.

Knijnik, G. (1996). *Exclução e resistência: Educação matemática e legitimidade cultural* [Exclusion and resistance: Mathematics education and culture legitimacy]. Porto Alegre, Brazil: Artes Médicas.

Knijnik, G. (2008). Educação matemática e diversidade cultural: Matemática cam-ponesa na luta pela terra [Mathematics education and cultural diversity: Farmworker mathematics in the struggle for land]. In P. Palhares (Ed.), *Et-nomatemática: Um olhar sobre a diversidade cultural e a aprendizagem matemática* (pp. 131–156). Ribeirão, Portugal: Edições Húmus.

Knijnik, G., Wanderer, F., & Oliveira, C. J. (Eds.) (2004). *Etnomatemática, currículo a formação de professores* [Ethnomathematics, a curriculum for teacher educa-tion]. Santa Cruz do sul, Brazil: EDUNISC.

Parra Sánchez, A. I. (2012). Etnomatemática e educação própria [Ethnomathema-tics and genuine education]. (Master's thesis). Rio Claro, Brazil: Instituto de Geociências e Ciências Exatas, Universidade Estatual Paulista (UNESP).

Searle, J. (1969). *Speech acts.* New York, NY: Cambridge University Press

Skovsmose, O. (2009). *In doubt: About language, mathematics, knowledge and life-world.* Rotterdam, the Netherlands: Sense.

Skovsmose, O. (2010). Symbolic power and mathematics. In R. Bhatia (Ed.), *Pro-ceedings of the International Congress of Mathematicians, Vol. 1* (pp. 690–705). Hy-derabad, India: Hindustan Book Agency.

Skovsmose, O. (2012a). Symbolic power, robotting, and surveilling. *Educational Studies in Mathematics, 80*(1), 119–132.

Skovsmose, O. (2012b). Mathematics as discourse. *Bolema* [*Mathematics Education Bulletin*], *26*(43), 1–18.

Sousa, F., Palhares, P., & Sarmento, M. (2008). Calafates na Baía de Câmara de Lobos [Caulkers in Baía de Câmara de Lobos.]. In P. Palhares (Ed.), *Etno-matemática: Um olhar sobre a diversidade cultural e a aprendizagem matemática* (pp. 157–191). Ribeirão, Portugal: Edições Húmus.

Taylor, F.W. (1911/2006). *The principles of scientific management.* New York, NY: Cosimo Classics.

Vierira, L., Palhares, P., & Sarmento, M. (2008). Etnomatemática: Estudo de elementos geométricos presentes na cestaria [Ethnomathematics: Study of geometrical elements present in basketry]. In P. Palhares (Ed.), *Etnomatemática: Um olhar sobre a diversidade cultural e a aprendizagem matemática* (pp. 291–315). Ribeirão, Portugal: Edições Húmus, 2008.

Whitehead, A. N., & Russell, B. (1910–1913). *Principia mathematica I-III.* Cambridge, England: Cambridge University Press

Wittgenstein, L. (1921/1992). *Tractatus logico-philosophicus* [Tractatus logico-philosophicus]. (C. K. Ogden, Trans.). London, England: Routledge.

Wittgenstein, L. (1953/1958). *Philosophical investigations* (2nd ed.). [G. E. M. Anscombe, Trans.]. Oxford, England: Basil Blackwell.

Žižek, S. (2008). *Violence.* New York, NY: Picador.

Ole Skovsmose
Aalborg University, Denmark, and
State University of São Paulo, Brazil

CHAPTER 10

EXAMINING POLITICAL PERSPECTIVES IN MATHEMATICS EDUCATION

Paola Valero and Alexandre Pais

MEETING EVA JABLONKA

After having seen her name in many papers in prominent mathematics education publications, the time had come to meet Eva Jablonka in person. For me (Paola), it was a late January evening in 2008, at the Swedish Mathematics Education Research Conference in frozen Stockholm. Jablonka had just accepted a professor's position at Luleå University; she was to meet the Swedish research community at the conference. It took a few sentences and a couple of cocktails before I was in the middle of a quite deep, critical—but at the same time sharply humorours—conversation with Jablonka. From our common interest in critical readings of mathematics education, to a shared experience of being a foreigner in Scandinavia, many topics have been good excuses to spend precious (short but quite exciting) time together since then.

Jablonka was for me (Alexandre), a risky figure. I cannot identify for sure where or when I first met her, but I remember well the intensity of her approach; and the effort I had to make to understand each of the 185 words she says per minute. The depth of her thought and her capacity to

Refractions of Mathematics Education, pages 173–195

articulate particular aspects of mathematics education with a strong political awareness, made some of the conversations we had a true joy. One can perk up more in a coffee break conversation with Jablonka, than by reading papers of the conference when a coffee break occurs. She does not bullshit, nor does she tolerate hypocrisy. To talk with her is to embark on a journey, where one has to be secure with one's own thoughts and positions. Do it differently, and you will become the target of Jablonka's surreptitious irony—without having a clue that she is mocking you.

For both of us talking to Jablonka has always been a challenge. Understanding her subtle yet sharp criticisms, often hidden in serious-sounding humorous comments, has led us to appreciate her as a person. As a colleague, whose work has constantly been in our conversations, we admire her scholarship; particularly in her call not to leave unattended the various myths that dress mathematics with a gown of neutrality.

Since her early years as a researcher Jablonka has always kept an eye on the societal aspects that delineate what we say and do in mathematics education. Her work has been known for criticizing taken-for-granted assumptions in mathematics education (for instance, concerning real-life problems, mathematical literacy and modeling) by way of highlighting how the surrounding culture is both influenced by and has influence on the particularities of mathematics education. This has been accomplished by a rigorous use of Basil Bernstein's work—a sociological theory at the margins of the traditional psychological approach to mathematics education—and a clear political stance toward the role played by school mathematics in current society. In particular, Jablonka's work has shown the fallacies of a certain naturalization perpetrated by research about the importance that mathematics has in people's mundane lives. In her writings with Uwe Gellert about a critical examination on the promises of mathematical modeling for primary and secondary school mathematics, a possibility was offered to demystify the relationship between mathematics and reality. The conceptualization of mathematical modeling as a generic competence became devoid of ethical stances toward the ideological purposes of the modelers. Thus, such generic competence could easily be contributed to a "sociomathematical indoctrination" (Jablonka & Gellert, 2011). Jablonka's work bears witness to the importance of de-naturalizing a certain "goodness" and neutrality intrinsic to a particular modern idea of mathematics. Thus, when doing research in mathematics education, mathematics is not enough: "[t]he ability to evaluate critically can neither be considered as mathematical nor automatically follows from a high level of mathematical knowledge" (Jablonka, 2003, p. 98).

In this chapter we will follow that path, and will pay our tribute to Jablonka by providing a succinct mapping of the way the political has been present in mathematics education research. We will finish by drawing on our own work to exemplify what can be seen and said about school mathematics, when we look at it from two different perspectives: Žižek's ideology critique, and Foucault's analytics of power.

THE POLITICAL DISPLACEMENT

A political perspective in mathematics education stands on the critical recognition that as soon as mathematical forms of knowing enter the scene of educational relationships, they stand on the dirty floor of humans with their entanglement in historical, social, economic, cultural, ethical, and political relationships. Mathematics education as a field of research has been built on the assumption of a necessary connection between the realm of mathematics as a field of research, and on mathematics education as the field of practice. In the realm of mathematics education, the historically produced knowledge by mathematicians has been reshaped in the contexts of education—schooling, adult education, informal education, etc. Such an assumption has been clearly identifiable in different theories of mathematics education. For example, Ernest (1998) in his socio-constructivist philosophy of mathematics and mathematics education, proposed that public mathematical knowledge, socially produced in the academic context of mathematics, entered the school context by a process of recontextualization. Such recontextualized knowledge was subject to personal reformulation in school learning and assessment.

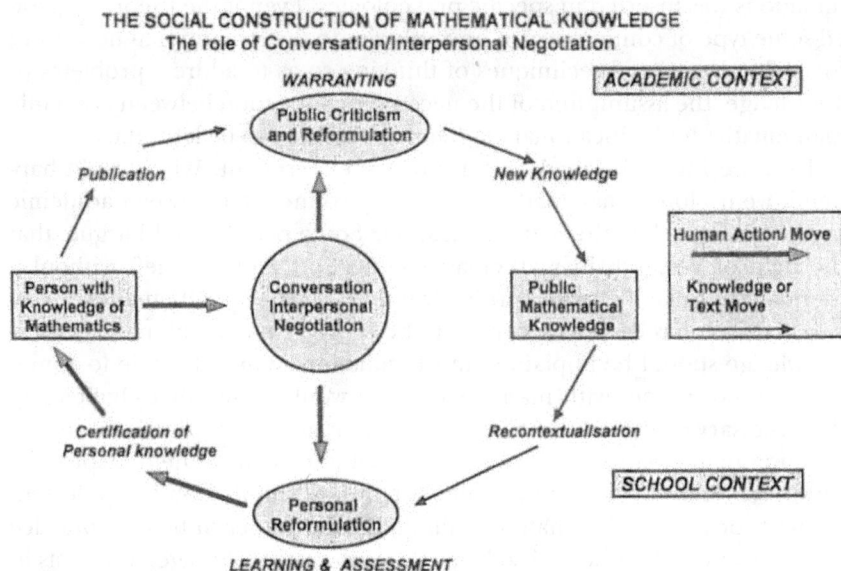

Figure 10.1 The creative/reproductive cycle of mathematics (Ernest, 1998).

Following this model, the way of reasoning about mathematics education would be expressed in statements such as: If academic mathematical knowledge Y is characterized by X, then the pedagogical or didactical question would be how a recontextualized knowledge Y' could be constructed or appropriated by students, so that both Y' and its recontextualized characteristics X' tend as much as possible to resemble Y and X. In other words, and put more simply, if mathematics is Y, then mathematics education should be Y'. The connection between Y and Y' as forms of knowledge has become a necessary connection since Y' is a recontextualized, simplified form derived from Y.

The French tradition of *didactique des mathèmatiques*, in its previous dominant theory of didactic transposition (Chevallard, 1985), and in its current form of the anthropological theory of the didactic (ATD) (Bosch and Gascón, 2006), also built on the assumption of the necessary link between the field of knowledge of academic mathematics and the transposed knowledge that becomes school mathematics. The theory of didactical transposition proposed a model to explain and describe the transformation of mathematical knowledge from its field of academic production into the field of schooling—or more broadly, of education—where it entered a didactic relationship between a learner and a person in charge of teaching it. While ATD attempted to emphasize the series of institutional (thus cultural and anthropological) determinations of the transposition, it was not clear how the mathematical knowledge that entered the systems of institutional determinations was inserted in specific praxeologies. Even if the theory suggested some type of contingency of praxeologies to the institutional necessities for technologies and techniques of thinking so as to address problems of knowledge, the assumption of the necessary connection between academic mathematics and educational mathematics seemed to be left intact.

Here we invite the reader to a thought experiment. What would happen if we no longer assumed the necessary connection between academic mathematics and mathematics education? Some people would argue, that the field of research in mathematics education would be left without a foundation, since the main (probably implicit) aim of mathematics education research has been to secure that the right foundational mathematical knowledge should be implanted in students for them to be able to appreciate, know and act with mathematics. We would argue that challenging the necessary connection between these two realms, is important to understanding mathematics education as a social and political field of practice within schooling. One could argue, borrowing Knijnik's philosophical argument for ethnomathematics (Knijnik, 2012), that each field of practice that we can call "mathematical," has particular rules of usage and exists in its particular form as it is shaped by the people who are part of such a practice. Using Wittgenstein's terminology, each of the fields would constitute a "language game." That we call different language games "mathematical," has to do with the fact that there are recognizable family resemblances between them. They are not identical; they are not the result of a transposition of one into the other. If we take this approach to understanding mathematics education as a field of the school curriculum, we should be asking questions about the constitution of that practice without considering that the mathematical content is the most salient and essential element of the field. Embedded in schooling, the organization of mathematics education has been constituted by the intermesh of historically formed practices of education; highly shaped in the 20th and 21st centuries by the political projects of national consolidation. It has been replacing religion-bound salvation narratives, with new scientific-based narratives of the literate, scientifically and mathematically capable citizen; and in growingly internationalized economic settings.

Such a way of understanding mathematics education leads us to a particular reading of the political. What if school mathematics was not important in society due to the exceptional and intrinsic characteristics of the academic field that gives this school subject its name; but rather due to the place it occupies within a particular social configuration of power? In the great majority of mathematics education research, certain features of mathematics (such as its neutrality or universality) are not perceived as politically marked, but appear as neutral, as part of a non-ideological common sense. But from a political perspective, it is precisely this neutralization of certain features within a naturalized background that is its most political characteristic. A political approach thus assumes that the teaching and learning of mathematics are not neutral practices, but that they insert people (be they children, youth, teachers, or adults) in socially valued mathematical rationalities and forms of knowing. Such an insertion is part of larger process of selection of people that schooling operates in society. It results in differential positioning of inclusion or exclusion of learners in relation to access to socially privileged resources, such as further education, the labor market, or cultural goods.

Seen this way, the specificity of mathematics and its teaching and learning comes not from its intrinsic characteristics, but from the sociopolitical importance assigned to this school subject in the historical process of schooling. A political approach to mathematics education aims at identifying a certain misrecognition that concerns the relationship between a structured network—the entire political world where mathematics education acquires its meaning—and one of its elements: mathematics itself. Depending on the kind of political and theoretical lens one uses, mathematics education can be understood in different ways. For instance, in our own work we have been using both the Foucauldian analytics of power and the Žižekian philosophy in order to posit mathematics within a given order or structure where its importance can be conceptualized in political terms (and not, as it often is, in terms of its immediate characteristics of problem solving, utility, beauty, or cultural possibilities). While one can use other theoretical grounds, the insight of a political approach to mathematics education concerns the idea, that if we want to understand mathematics education and how it effects power and is effected through power, we have to step out from it; we have to posit it in an entanglement within broader societal relationships. For the two writers of this chapter, such a research strategy to approaching mathematics education is a common interest. However, the theoretical elements that each has chosen to play with in such an endeavor, allows us sometimes to talk in unison and other times to differentiate our voices. For Paola, drawing on Foucaultian cultural historical studies on power, mathematics education in the school curriculum is a site for the study of the political constitution of modern subjectivities. While for Alexandre, drawing on Žižek's perspective, the structure that colors mathematics education and gives it its ultimate meaning, is the economical and ideological constitution of capitalism. Yet, in either case, the political move concerns a displacement from analyzing the direct properties of the mathematical element for the purpose of formulating recommendations for teaching and learning practices, to an analysis of the network of social and political relations of which mathematics is an effect. Such an analysis is not directly concerned with generating knowledge for improving the didactics—in the European sense of the word—of school mathematics; it rather poses a challenge to research itself, by showing how the workings of mathematics education research partakes in particular systems of reason, and in ideologies that format what is possible to think about practice.

TRACING THE EMERGENCE AND MAPPING THE PRESENT

The political perspectives of mathematics education became a concern for teachers and researchers in the 1980s. While the change from the 19th century to the 20th century was a time when the inclusion of mathematics as a school subject was growing massive, national education systems around the world (Clements, Keitel, Bishop, Kilpatrick, & Leung, 2013); the change from the 20th century to the 21st century has been a time for focusing on the justifications for the privileged and expanding role of mathematics in educational systems at all levels. The apparent failure of the new-math movement in different industrialized countries, called attention to the importance of a quality mathematical instruction that could reach as many students as possible and not just a select few (Damerow, Dunkley, Nebres, & Werry, 1984). The issue emerged of how a mathematics education, which had focused narrowly on the fundamental concepts and structures of mathematics, had resulted in an elitist education; an education that had little potential for reaching the masses. The growing attention to the "unexpected, side effects" of the new mathematics curriculum, fed the political concern and involvement of many mathematics educators. Such concern became an initial entry that allowed sensitivity and awareness about researching how mathematics education could be "political" (Lerman, 2000). Such political awareness on issues, such as how mathematics has played a role in society as a gatekeeper to entry into further education, has been important for many researchers whose work has shaped the role of political perspectives in mathematics education.

Questions started to be raised in the 1980s, about how mathematics education could be studied from perspectives that allowed moving beyond the boundaries of the mathematical contents in the school curriculum. In mathematics education, the first book published in English as part of an international collection containing the word *politics* in the title was *The Politics of Mathematics Education* by Stieg Mellin-Olsen (1987), who was a Norwegian researcher and teacher educator at Bergen Teacher College. Mellin-Olsen's book traveled through the fields of psychology, anthropology, psychoanalysis, and sociology; it showed their relevance to research and used them to delineate what could be a political approach to mathematics education. However, and notwithstanding Mellin-Olsen's efforts to politicize school mathematics (to understand it as an element of a political structure), the interest driving his attempt to broaden the scope of his thinking was still a didactical one. Mellin-Olsen was more interested in perceiving the political factors influencing or inhibiting a meaningful mathematics learning, so that they could be overcome for the sake of the teaching and learning of mathematics. He did not seem to be interested in criticizing the political role played by school mathematics in the differentiation of students.

In that same decade, the fields of critical mathematics education and ethnomathematics emerged and brought into the field of mathematics education cultural, social, and political aspects of mathematics that are now widely recognized as influential when learning this school subject. The work of Marilyn Frankenstein and Ole Skovsmose can arguably be considered the seminal and most influential research within critical mathematics education. The conceptualization of critique in the work of Frankenstein was grounded on Freire's liberatory pedagogy, where the notions of *conscientização* (critical consciousness) and *transformation* were crucial to the thinking about educational practice (Freire, 1998). Her work has influenced new generations of researchers who also take advantage of Freire's critical pedagogy (Gutstein, 2003). On the other hand, Skovsmose (1994) understood critical education as one that addressed the conflicts and crises in society by uncovering inequalities and oppression of whatever kind (p. 22). Addressing the critical role played by mathematics in society, implied an understanding of the risks and uncertainties that mathematics and societal progress conveyed. In the field of mathematics education, a critical approach could involve confronting students with situations in which mathematics seemed to format the way they understood and acted upon reality. Within the theoretical frameworks informing Skovsmose's and Frankenstein's work, there is, despite the differences, a strong affinity. The presence of Freire's theory in Skovsmose's notions of *mathemacy* and "dialogic learning and teaching" (Alrø & Skovsmose, 2002), as well as the fact that Freire himself was informed by theories coming from the Frankfurt School of critical theory, are probably the most visible similarities between these two approaches to critical mathematics education. From the authors' points of view, both approaches are open for questioning the purity and neutrality assumed in both mathematics and mathematics education. However, both authors have concern with a research that has as an ultimate aim, to inspire better educational practices of teachers and in classrooms.

Despite being often confused with an ethnic or indigenous mathematics, ethnomathematics does not restrict its research to the mathematical knowledge of culturally distinct people, or people in their daily activities. The focus of ethnomathematics could be academic mathematics, through a social, historical, political, and economical analysis of how mathematics has become the privileged knowledge it is today. Within this background, ethnomathematics research has brought to the mathematics education field new and refreshing insights, not just about ethnic or local mathematical knowledge, but also about philosophical, historical, and political approaches involved in mathematics and its education. Ubiratan D'Ambrosio (2002) defines ethnomathematics as, "a research program in the history and philosophy of mathematics, with obvious pedagogic implications" (p. 27); and he points out some of the dimensions involved in the ethnomathematical research: historical, cognitive, epistemological, political, and educational. Despite this all-encompassing goal, research in ethnomathematics has become predominantly focused on local cultures and non-scholarly forms of mathematics (Pais, 2011). The premise justifying this kind of research goes as follows: students already have some kind of pre-school, proto-mathematical knowledge; this knowledge should be considered by the teacher when organizing the learning of school mathematics; this way it is assured that cultural differences are valorized, and a better learning can occur (since students do not start from scratch, but from their own life experiences). In this perspective, ethnomathematics becomes another of the many didactical tools that abound in research. That is, it stops to be a critical reflection on the sociopolitical roots of academic mathematics, and instead the place that it occupies in the social imaginary and in schooling has become as a learning device.

The previous authors who clearly formulated initial political perspectives in mathematics education, were mathematics educators. Nevertheless there were other researchers who put their gaze on the field. Valerie Walkerdine (1988) argued that what appeared as irrational, was positioned as a threat to the "power of scientific rationality, a power clearly related to the rise of the bourgeoisie" (p. 60). A great deal was invested in the production of a modern discourse based on the idea of the subject as a rational and conscious one. All the subjectivities that somehow did not match this ideal (the feminine, the mob, the proletariat, the savage), were posited as unreasonable, therefore as a threat to be eliminated. Through her analysis of school mathematics practices, she showed how school mathematics teaching and learning practices inscribed in children specific notions of abstract thinking and of the rational child. Such notions were already being produced in society, and particularly inside the scientific educational discourse, but they became a practice that fabricated subjects within the everyday routines of mathematics classrooms. Walkerdine's work has become a cornerstone for future uses of poststructuralist theories in mathematics education.

In the last two decades, we have witnessed a dissemination of the research initiated with the work of these scholars, and a growth of political perspectives concerning mathematics education and democracy, equity, and social justice. A considerable amount of journal special issues and edited collections, have gathered a number of papers reporting theoretical discussions and empirical work within this line; not to mention that in the *Third International Handbook of Mathematics Education* (Clements et al., 2013) a whole section called "Social, Political and Cultural Dimensions in Mathematics Education" included eight papers on the topic, out of which three had a clear political perspective. In general terms, the growing amount of research has provided an initial positioning of mathematics and school mathematics within the social and political spectrum of our times. There is still at the core of this research a desire for the betterment of teaching practices, since most of the research adhered to the ground assumption defining mathematics education research: that research has to be connected with the problems of practice. This demarcation of a disciplinary boundary in the field, has been productive but has also limited the possibilities for understanding the political in mathematics education (Pais and Valero, 2011). There has also been a move from the naïve modern idea of utopian progress, in which human beings through the deployment of their mathematical rationality will achieve the ideal high-tech society with plenty of resources available to anyone; toward more localized concerns with empowerment and liberation from a plurality of social constraints and injustices associated with race, sexuality, ecology, language, cultural minorities, colonialism, religion, rurality, and class (D'Ambrosio, 2002; Ernest, 2004;

Gutstein, 2003; Gutiérrez, 2010; Knijnik, 2007; Valero & Stentoft, 2010). Nowadays the spectrum of work that could be mapped as political is broad.

Despite the expansion of the area, it is important to raise an analytical distinction. Although the political awareness of researchers has been a point of departure for this type of research (as Lerman (2000) has suggested), it does not constitute the center of a political approach. It is different to be sympathetic to how mathematics education relates to political processes of different type, than to make the study of power in mathematics education the analytical focus of one's research. In other words, not all people who express a political sympathy and whose papers touch on the politics of mathematics education, actually study the political in mathematics education (Gutierrez 2010; Valero 2004). The latter calls the researcher to deploy analytical strategies and conceptualizations where power is central. With this distinction in mind, it is possible to differentiate a variety of political perspectives. Some could be called weak, in the sense that they make a connection between mathematics education and power; but they do not concentrate on the study of power as a constituent of mathematics education, but rather as a result or a simply associated factor. Strong political approaches in mathematics education have a variety of perspectives, which do have an explicit interest in understanding mathematics education as political practices. In what follows we will map a diversity of work in these two directions.

THE POLITICS OF MATHEMATICS EDUCATION

To use the adjective *weak* to categorize a colleague's work, is a quite contentious move. In this section, rather that placing particular names and references, our intention will be to point in general to the types of ideas and arguments that circulate in this type of research.

A first general idea of weak political perspectives in mathematics education is that the political in mathematics education, is present in external political and economic factors that influence the teaching and learning of mathematics. Such factors are important to identify. However they are simply external conditions surrounding the work of teachers and learners in mathematics classrooms. Examples are studies of how mathematics education practices are shaped by educational policies. Many studies have discussed the influence of national or local reforms in shaping mathematics curricula around the world. A paradigmatic case could be South Africa. Given the transition from apartheid to democracy at the beginning of the 1990s, there has been a proliferation of studies showing how deeply the policy changes influenced how and why mathematics education in primary and secondary school was transformed to contribute—or not—to the construction of a new democratic society. Many of these studies operate on some political assumptions on mathematics education, and on its role in society; but they are intend to study appropriate pedagogies, and not how pedagogies in themselves affected the exclusion that the programs intended to remediate.

A second characteristic of this type of studies, is the adherence to some of the positive features attributed to mathematics and mathematics education; particularly those that have to do with people's empowerment through mathematics (learning). More often than not, it is assumed there is some kind of intrinsic goodness in mathematics and mathematics education (Skovsmose & Valero, 2001), which is transferred from teachers to learners through good and appropriate education practices. The discourse of the power of mathematics and mathematics education to achieve people's empowerment, has been built on the assumption that mathematics, being the purest and most abstract form of human thinking, develops not only mental capacities and structures; but also, as a consequence, people's competencies for action. Once mental structures are in place, individuals are capable of participating in actions; such as describe, count, measure, control, predict, argue, communicate, model, criticize, and even transform their social and material world. A person who has learned mathematics (properly, of course), has gained these possibilities from having seen the world with mathematics. Mathematics teachers, the possessors of such knowledge and of those capacities, transfer mathematics to students. Students then become empowered by the acquisition of a knowledge, which allows them to exercise powerful actions. Even if this description of the reasoning about mathematical empowerment seems to misrepresent the complex ideas or that appear to justify mathematics education practices and mathematics education research, the bone of such thinking seems to have become a commonsense idea about the political dimension of mathematics education, to which a large amount of research has alluded.

In the 1980s and into the 1990s, the broadening of views on mathematics education allowed for formulations of the aims of school mathematics in response to social challenges of changing societies, in particular, in response to the consolidation of democracy. It became possible to enunciate the idea that (as part of a global policy of education for all by UNESCO), mathematics education had to contribute to the democratic competence of citizens, and also to open access for all students. In many countries, both at the national-policy level and at the level of researchers and teachers, there was a growing concern about mathematics for all and mathematics for equity and inclusion (Jurdak, 2009). The study of how different groups —women, linguistic, ethnic or religious minorities, particular racial groups, etc.—of students systematically underachieved and how to remediate that situation, grew extensively. The interest of bringing mathematical empowerment to marginalized groups was, at that time, located within the discourses of education for democratization.

Finally, a third idea was the necessity of high mathematical achievement for economic growth and social progress. The discourses of education for democracy had progressively been displaced by those of strengthening a global agenda of education; and thus for a productive work force supported by the economic view of education by the OECD. Since the the early 2000s, mathematics (and nowadays science) had turned into the privileged school subjects to realize the promise of a population trained to be the productive force of a global, capitalistic, and competitive economy. In this new type of discursive configuration, "no child should be left behind" mathematically. Thus, the issues of access to and success in mathematics, had become economic imperatives for national governments as well as for mathematics education researchers. High mathematics achievement, thanks to the raise and influence of international comparative studies, had become more than ever a part of the promise of individual, community, and national growth, well-being and prosperity. Although this idea was not new in mathematics education, and has been present since the turn of the 19th century to the 20th century, it has become stronger than ever.

The Analysis of Power at the Centre

Other studies in mathematics education problematize the assumed neutrality of mathematical knowledge, and provide new interpretations of mathematics education as practices of power within the larger framework of social institutions.

Some of the work in ethnomathematics has developed an epistemological critique on the enduring belief in the universality and neutrality of mathematical knowledge; thus it challenged the supremacy of Eurocentric understandings of mathematics and mathematical practices. Some of these studies have brought a strong cultural critique to the imperialism of western culture, in defining what mathematics is meant to be. In this value ranking of knowledge, the technical knowledge and related practices of non-western communities have been deemed epistemologically inferior. Several studies have argued that epistemology was less a discussion on whether knowledge is itself universal or local, but what intentions and what politics allow us to claim that some knowledge (like academic mathematics) is universal (Powell & Frankenstein, 1997; Knijnik, 1998, 2007). As such, an analysis of power in ethnomathematics has been presented in studies that not only argued for how the mathematical practices of different cultural groups—not only indigenous or ethnic groups, but also professional groups—were of epistemological importance and value; but also how some of those cultural practices have been inserted in the calculations of power, so that they could construct a regime of truth around themselves, and thus gain a privileged positioning in front of other practices (Knijnik, 2012).

Critical mathematics education took a point of departure in a variety of theories. Some scholars and practitioners (Frankenstein, 1995) have used Marxism's ideas concerning education and ideology to study how mathematics education has been implicated in the process of exclusion and differentiation of students, when mathematics education practices reproduce class position and student disadvantage. Taking elements from the philosophy of science, the Frankfurt School's critical theory, and Freirian critical education, other scholars (Skovsmose, 1994) have raised a double critique. On the one hand, it has been important to study how mathematics is a formatting power for technological, scientific, and social action through its use in the creation of scientific and technological structures that operate in society (Christensen, Skovsmose, & Yasukawa, 2008). But on the other hand, the study of mathematics education practices has shown how teaching and learning situations can be critical (Skovsmose & Valero, 2008); that is, they can be either constructive and promote reflections and democratic competence that lead to a critical stance toward the uses of mathematics in society; or they can be differentiating, segregating, and oppressive for those who do not succeed in them. Critical mathematics education research has provided both frameworks for understanding and studying the link between mathematics education and power in society; and has created the frameworks to think about the transformation of educational practice toward the fulfillment of goals, such as empowerment and emancipation (Gutstein, 2003).

The study of the political in relation to the alignment of mathematics education practices with Capitalism, has also been a recent and strong political reading of mathematics education, which offers a critical perspective on the material, economic significance of success in mathematics education. Both educational practices (Baldino & Cabral, 2006) and research practices, lock students into a credit system where success in mathematics represents value. For instance, Sverker Lundin (2012), after analyzing how word problems were researched in mathematics education, concluded that the usefulness of mathematics for solving real-life problems was not a consequence of any direct properties of this science; but it was from the results from the workings of mathematics education itself. Moreover, the symbolic discourse around the importance of mathematics for everyday activities, has concealed the real importance of mathematics as a testing and grading device. What is seen as a direct property of mathematics (its utility), has indeed been the result of the place it occupies within the structure of capitalist economics.

In the United States, and as a reaction to the endemic operation of race as a strong element in the classification of people's access to cultural and economic resources, the recontextualization of critical race theories into mathematics education has provided new understanding of mathematics education as a particular instance of a white dominant institutional space that excludes educational success for African American learners (Martin, 2011). Martin has argued that operating in mathematics education research with such a sociological approach

> calls not only for an examination on how race and racism structure the very nature of the mathematics education enterprise but also for an examination of how mathematics education research, policy, and reform contribute to the dynamics of race and racism in the larger society. The goal of such an analysis is not to focus on individuals within the domain but rather on the configurations of power and practice, including discursive practices, which necessarily link mathematics education to racialized structures, process, and agendas in the rest of society. (Martin, 2011, p. 437)

In the case of Latinas (Gutiérrez, 2010), critical Latino studies and theories have helped focus on how mathematics education practices in classrooms, schools, and communities relate to racism and other forms of subordination: this through sexuality, language, immigration status, and phenotype (Gutiérrez, p. 6). These studies argued for a broader understanding of the issues of systematic differential access of Latina students as expressed in the historical achievement gap between different racial groups in the United States. Conceiving of mathematics education as complex educational political problems, allows to understand how

> mathematics has an influence not just on our educational experiences and identities within classrooms and schools, but mathematics also formats our lives by providing a lens onto our worlds. It is only when we are able to question the practices that occur within schooling, as well as those that operate beyond it, and in conducting mathematics education research that we can really begin to address a broader sense of equity (and transformation) in society. (Gutierrez, 2012, p. 33)

The re-contextualization of post-structural theories in mathematics education has also led to the study of power in relationship to the historical construction of modern subjectivities. The effects of power on the bodies and minds of students and teachers (Walshaw, 2010), as well as the public discourses on mathematics (Moreau, Mendick, & Epstein, 2010); have been studied in an attempt to provide insights into how the mathematical rationality that is at the core of different technologies in society, shape the meeting between individuals and their culture. Even though most research has concentrated on the issue of identity construction and subjectivity, some studies have attempted cultural histories of mathematics as part of a modern, massive educational systems; these are also broadening this type of political perspective (Knijnik & Wanderer, 2010; Popkewitz, 2004; Valero, García, Camelo, Mancera, & Romero, 2012).

OPERATING IN THE POLITICAL: A DIFFERENT READING OF THE IMPORTANCE OF MATHEMATICS

In what follows we will draw two different analyses of mathematics taken not in itself (as knowledge or competence); but on one hand, as part of a political economy, and on the other hand, as a crucial element for the constitution of modern subjectivity. These are only sketches of the research we have been conducting, and they intend to exemplify what can be a political perspective on mathematics education.

The Fetish of Mathematics, by Alexandre

It was Karl Marx (1867/1967) who first elaborated on the social and political implications of the mismatch between structure and its elements, apropos *commodity fetishism*. The essential feature of commodity fetishism was a certain misrecognition that concerned the relation between a structured network and one of its elements:

> What is really a structural effect, an effect of the network of relations between elements, appears as an immediate property of one of the elements, as if this property also belongs to it outside its relation with other elements. (Žižek, 2008, p. 19)

An emblematic example was given by Marx himself: "[f]or instance, one man is a king only because other men stand in the relation of subjects to him. They, on the contrary, imagine that they are subjects because he is a king" (p. 19). In other words, people think they treat the king as a king, because he is in himself a king; but in reality, a king is a king because we treat him like one. That is, the kingship of a king is strictly determined by the crowd's belief in it. In the last instance, it is not the crowd that is determined by the king (by his rules or commandments), but the king himself who depends on the belief of the crowd.

And cannot the same be said apropos mathematics? We think that we treat mathematics as an important and powerful knowledge because it is in itself a powerful knowledge; but in reality mathematics, is powerful because we treat it as such. However, as Žižek (2008, p. 161) has alerted us, the fact that the charismatic power of a king is an effect of the symbolic ritual performed by his subjects must remain hidden: as subjects, we are necessarily victims of the illusion that the king is already in himself a king.

And again, the same with mathematics: in order for mathematics to be perceived as a powerful knowledge, we must be blind to the fact that it is only an effect of our positioning of it as such.

Within Marxian theory, a commodity can take two values: use value and exchange value. While the use value of a commodity is strictly related to the concrete use someone makes of a commodity (the mathematical know-how necessary to perform a profession, for instance) the exchange value posits this commodity in relation to all the others; that is, as part of a structure of equivalences where its value can be gauged. Thus exchange value has a purely relational status: it is not inherent to a commodity. It expresses the way this commodity relates to all the others. When looking at a commodity, say a table, we see its use value. What we cannot see is its exchange value, which remains invisible.

If we transpose this line of thought to school mathematics, we can speculate how the gesture of positing the value of mathematics in its use, hides its exchange value; that is, the formal place school mathematics occupies within capitalism. The fundamental gap between an object and the structural place it occupies, is at work in school mathematics: while perceiving the importance of school mathematics as knowledge or competence, we neglect the importance of this subject in maintaining a schools' functioning as credit systems (Baldino & Cabral, 1998; Vinner, 1997). Such obliteration of the economical role of school mathematics, is a characteristic of research (Pais, 2013); and one that can be argued so as to make research innocuous, if the purpose is to change the worldwide problem of failure in this school subject.

The Making of the Mathematical Child, by Paola

Only a heretic would remove mathematics as a subject in the school curriculum, or reduce the amount of hours for it. The privileged position of school mathematics in current curricula is beyond any doubt. It has become an unquestionable truth of educational practices and research. But it does not take more than looking at educational practices a bit more than a century ago to realize that arithmetic—not even mathematics—was hardly a subject in school curricula (Howson, 1974). In many European countries, the 19th century was a time of construction of schools for the majority of the population. The dominance of the classic school with languages, particularly Latin, as the privileged subject, started being transformed for the purpose of providing an education suitable for the practical training of a workforce. In Luxembourg, for example, arithmetic was associated with subjects connected to accounting and economics, and it was made part of the practical subjects (Schreiber, 2014). In Denmark, it was only in 1903 where mathematics became a subject in the public school curriculum. The argument of arithmetic as a tool to sharpen the mind and to acquire practical skills, was put forward as a justification for the creation of the subject in the first years of primary school and lower secondary (Hansen et al., 2008, pp. 33–34). During the 20th century, new configurations of the school subject took place within the expanding education systems, and their project of consolidation of nation states. A little historization would make clear that the privilege of school mathematics is a contingent historical event. The interesting question emerging from a cultural historization of school mathematics is: which historical conditions have made possible the current narrative about the necessary dominance of mathematics in the school curriculum?

This question has been addressed by some of the people who research on the history of mathematics education and the school curriculum. Most of these accounts, however, have recurrently relied on the three arguments. Niss (1996, pp. 14–15) argues that:

> Mathematics education can indeed contribute to the technological and socio-economic development of society at large, . . . to society's political, ideological and cultural maintenance and development, . . . and to providing individuals with prerequisites which may help them to cope with life in the various spheres in which they live.

These three arguments are related to the intrinsic assumptions about the power of mathematics in relation to technological and social development, to culture creation, and to individual action. But what if we operated the displacement of the intrinsic assumptions of the power of mathematics, and thought of power and mathematics education in a different way?

If one thinks of mathematics education practices in schools as an area of the school curriculum, from the point of view of cultural historical studies in education grounded on Foucault and other contemporary social and cultural studies (Popkewitz, Franklin, & Pereyra, 2001); then there is no reason to suppose that the significance of school mathematics is a function of the internal characteristics of mathematics, when it is brought into the realm of schooling. Rather, it is the making of the desirable, rational and modern citizens, whose conduct is governed through schooling that makes mathematics education practices a site of power. These series of practices, from the curricular policy to the small didactical realizations of teachers in the classroom; they all inscribe on children's minds, bodies, and subjectivities not only a given knowledge, but also a whole system of reason with its associated epistemologies. Mathematics education practices then become a space of cultural politics, where ideas of the reasonable child are constantly made possible.

As an example, Valero et al. (2012) examined how school mathematics in Colombia intersected with the project of Spanish colonization, and the subsequent attempts at modernizing the state. The curricular technologies implemented massively in the 1980s, which were based on the expert knowledge of Piagetian cognitive theory and mathematical systemic thinking, installed a dispositive that should have secured the making of the rational modern child. Such an attempt, while forming and including certain subjectivities, operated also a differentiation of other forms of subjectivity, which did not match with the norm established by the curriculum. In this way, as technologies of government, the school mathematics curriculum affected inclusion and exclusion of children in the norm of the desired citizen.

LOOKING FORWARD

As researchers have engaged in making mathematics more meaningful and pleasurable, we have passed over questions whose answers are normally taken for granted. One of these questions is quite an elementary one: why is (school) mathematics important? The reasons may vary depending on one's area of research. For some researchers, mathematics is important because it contributes to the development of higher psychological functions: logical thinking, abstraction, metacognition, or creativity (Stech, 2008). For other, the importance of learning mathematics has more to do with the acquisition of mathematical instruments for solving everyday problems (de Lange, 1996); whereas others privilege a hedonistic dimension, by emphasizing how aesthetic, pleasurable, and attractive the learning of mathematics can be (Boaler, 2009). More recently there has been a growing emphasis on the cultural importance of mathematics, with researchers stressing the range of possibilities that mathematics offers people in having a social, historical, and cultural experience (Radford, 2012). On a more critical note, some researchers have emphasized how powerful mathematics can be in formatting reality, which requires an exploration of real mathematical models in a critical way (Skovsmose, 1994). Although these five areas of importance can be categorized into more nuanced goals of mathematics teaching (Niss, 2007; Bishop & Forgasz, 2007), we can see how the importance of mathematics has invariably been located in the immediate properties of mathematics. That is, the reasons invoked to justify the importance of mathematics in schools have been conceived in terms of its *inherent* characteristics, whether they are related to the development of mental functions, the utility of this school subject for people's lives, its beauty, cultural richness, or the ideals of citizenship. Mathematics seems to embody the right properties that make it important.

Against this perspective, we argue that the crucial step for a political approach in mathematics education should imply a displacement of the way the importance of this school subject is conceived. Instead of conceiving the importance of mathematics in terms of mathematics itself, we conceive its importance in terms of the place this subject occupies within a given societal arrangement; in our case, we then theorize in terms of Foucault's analytics of power or Žižek's ideology critique. That is, we see mathematics as important not because of its intrinsic characteristics transferred to school mathematics (problem-solving, utility, beauty, cultural possibilities, etc.), but because of the sociopolitical importance assigned to it. Such a step is a difficult one to give within a field historically centered in mathematics. It may well imply a complete redefinition of the way in which mathematics education is perceived by those who work in it; but this is the precise purpose of a sociopolitical approach to mathematics education (Gutiérrez, 2010, p. 20; Sriraman & English, 2010, p. 25).

ACKNOWLEDGEMENTS

This paper elaborates further on the ideas in Valero (2014).

REFERENCES

Alrø, H., & Skovsmose, O. (2002). *Dialogue and learning in mathematics education: Intention, reflection, critique.* Dordrecht, the Netherlands: Kluwer.

Baldino, R., & Cabral, T. (1998). Lacan and the school's credit system. In A. Olivier & K. Newstead (Eds.), *Proceedings of 22nd conference of the international group for the Psychology of Mathematics Education (PME22), Vol. 2* (pp. 56–63). Stellenbosch, South Africa: University of Stellenbosch.

Baldino, R., & Cabral, T. (2006). Inclusion and diversity from Hegel-Lacan point of view: Do we desire our desire for change? *International Journal of Science and Mathematics Education, 4,* 19–43.

Bishop, A., & Forgasz, H. (2007). Issues in access and equity in mathematics education. In F. Lester (Ed.), *Second handbook of research on mathematics teaching and learning* (pp. 1145–1168). Charlotte, NC: Information Age.

Boaler, J. (2009). *The elephant in the classroom: Helping children learn and love maths.* London, England: Souvenir Press.

Bosch, M., & Gascón, J. (2006). Twenty-five years of the didactic transposition. *ICMI Bulletin, 58,* 51–63.

Chevallard, Y. (1985). *La transposition didactique* [The didactic transposition]. Grenoble, France: La Pensée Sauvage.

Clements, M. A., Keitel, C., Bishop, A., Kilpatrick, J., & Leung, F. S. (2013). From the few to the many: Historical perspectives on who should learn mathematics. In M. A. Clements, A. J. Bishop, C. Keitel, J. Kilpatrick & F. K. S. Leung (Eds.), *Third international handbook of mathematics education* (Vol. 27, pp. 7–40). New York, NY: Springer.

Christensen, O. R., Skovsmose, O., & Yasukawa, K. (2008). The mathematical state of the world – Explorations into the characteristics of mathematical descriptions. *Revista de Educação em Ciência e Tecnologia 1*, 77–90.

D'Ambrosio, U. (2002). *Etnomatemática: Elo entre as tradições e a modernidade* [Ethnomathematics: Linking tradition with modernity]. Belo Horizonte, Brazil: Autêntica.

Damerow, P., Dunkley, M., Nebres. B., & Werry, B. (Eds.) (1984). *Mathematics for all.* Paris, France: UNESCO.

De Lange, J. (1996). Using and applying mathematics in education. In A. Bishop, M. Clements, C. Keitel, J. Kilpatrick, & C. Laborde (Eds.), *International handbook of mathematics education* (pp. 49–97). Dordrecht, the Netherlands: Kluwer.

Ernest, P. (1998). *Social constructivism as a philosophy of mathematics.* Albany, NY: State University of New York Press.

Ernest, P. (2004). Postmodernism and the subject of mathematics. In M. Walshaw (Ed.), *Mathematics education within the postmodern* (pp. 15–33). Greenwich, CT: Information Age.

Frankenstein, M. (1995). Equity in mathematics education: Class in the world outside the class. In E. Fennema & L. Adajian (Eds.), *New directions for equity in mathematics education* (pp. 165–190). Cambridge, England: Cambridge University.

Freire, P. (1998). Cultural action for freedom. *Harvard Educational Review, 8*(4), 471–521.

Gutiérrez, R. (2010). The sociopolitical turn in mathematics education. *Journal for Research in Mathematics Education, 41*(0), 1–32.

Gutiérrez, R. (2012). Context matters: How should we conceptualize equity in mathematics education? In B. Herbel-Eisenmann, J. Choppin, D. Wagner, & D. Pimm (Eds.), *Equity in discourse for mathematics education* (pp. 17–33). Dordrecht, the Netherlands: Springer.

Gutstein, E. (2003). Teaching and learning mathematics for social justice in an urban, Latino school. *Journal for Research in Mathematics Education, 23*(1), 37–73.

Hansen, H. C., Haar, O., Jensen, H. N., Wedege, T., Jakobsen, I. T., Thybo, C., . . . Jørsboe, O.G. (2008). *Matematikundervisningen i Danmark i 1900-tallet. Grundlæggende regning og matematik* [Mathematics teaching in Denmark in the 1900s. Basic arithmetic and mathematics] (Vol. 1). Odense, Denmark: Syddansk Universitetsforlag.

Howson, G. (1974). Mathematics: The fight for recognition. *Mathematics in School, 3*(6), 7–9.

Jablonka, E. (2003). Mathematical literacy. In A. Bishop, M. A. Clements, C. Keitel, J. Kilpatrick, & F. K. S. Leung (Eds.), *Second international handbook of mathematics education* (Vol. 1, pp. 75–102). Dordrecht, the Netherlands: Kluwer.

Jablonka, E., & Gellert, U. (2011). Equity concerns about mathematical modeling. In B. Atweh, M. Graven, W. Secada, & P. Valero (Eds.), *Mapping equity and quality in mathematics education* (pp. 223–236). New York, NY: Springer.

Jurdak, M. (2009). *Toward equity in quality in mathematics education.* New York, NY: Springer.

Knijnik, G. (1998). Ethnomathematics and postmodern thinking: Convergences/divergences. In P. Gates (Ed.), *Proceedings of the first international conference on mathematics education and society (MES1)* (pp. 248–252). Nottingham, England: Centre for the Study of Mathematics Education.

Knijnik, G. (2007). Mathematics education and the Brazilian landless movement: Three different mathematics in the context of the struggle for social justice. *Philosophy of Mathematics Education Journal (Online), 21.*

Knijnik, G. (2012). Differentially positioned language games: Ethnomathematics from a philosophical perspective. *Educational Studies in Mathematics, 80*(1), 87–100.

Knijnik, G., & Wanderer, F. (2010). Mathematics education and differential inclusion: A study about two Brazilian time–space forms of life. *ZDM–The International Journal on Mathematics Education, 42*(3–4), 349–360.

Lerman, S. (2000). The social turn in mathematics education research. In J. Boaler (Ed.), *Multiple perspectives on mathematics teaching and learning* (pp. 19–44). New York, NY: Ablex Publishing.

Lerman, S. (2013) (Ed.), *Encyclopedia of Mathematics Education.* New York, NY: Springer Reference. Retrieved from http://www.springerreference.com/docs/edit/chapterdbid/313305.html

Lundin, S. (2012). Hating school, loving mathematics: On the ideological function of critique and reform in mathematics education. *Educational Studies in Mathematics, 80,* 73–85.

Martin, D.B. (2011), What does quality mean in the context of white institutional space? In B. Atweh, M. Graven, W. Secada, & P. Valero (Eds.), *Mapping equity and quality in mathematics education* (pp. 437–450). New York, NY: Springer.

Marx, K. (1867/1967) Engels, F. (Ed.). *Capital: A critique of political economy,* (Vol. 1). New York, NY: International Publishers.

Mellin-Olsen, S. (1987). *The politics of mathematics education.* Dordrecht, the Netherlands: Kluwer.

Moreau, M. P., Mendick, H., & Epstein, D. (2010). Constructions of mathematicians in popular culture and learners' narratives: A study of mathematical and non-mathematical subjectivities. *Cambridge Journal of Education, 40,* 25–38.

Niss, M. (1996). Goals of mathematics teaching. In A. J. Bishop, K. Clements, C. Keitel, J. Kilpatrick, & C. Laborde (Eds.), *International handbook of mathematics education* (Vol. 1, pp. 11–47). Dordrecht, the Netherlands: Kluwer.

Niss, M. (2007). Reflections in the state and trends in research on mathematics teaching and learning: From here to utopia. In F. Lester (Ed.), *Second handbook of research on mathematics teaching and learning* (pp. 1293–1312). Charlotte, NC: Information Age.

Pais, A. (2011). Criticisms and contradictions of ethnomathematics. *Educational Studies in Mathematics, 76*(2), 209–230.

Pais, A. (2013). An ideology critique of the use-value of mathematics. *Educational Studies in Mathematics, 84*(1), 15–34.

Pais, A., & Valero, P. (2011). Beyond disavowing the politics of equity and quality in mathematics education. In B. Atweh, M. Graven, W. Secada, & P. Valero (Eds.), *Mapping equity and quality in mathematics education* (pp. 35–48). New York, NY: Springer.

Popkewitz, T. S. (2004). The alchemy of the mathematics curriculum: Inscriptions and the fabrication of the child. *American Educational Research Journal, 41,* 3–34.

Popkewitz, T. S., Franklin, B. M., & Pereyra, M. A. (2001). *Cultural history and education: critical essays on knowledge and schooling.* New York, NY: RoutledgeFalmer.

Powell, A., & Frankenstein, M. (1997). *Ethnomathematics: Challenging eurocentrism in mathematics education.* Albany, NY: State University of New York Press.

Radford, L. (2012). Education and the illusions of emancipation. *Educational Studies in Mathematics, 80*(1–2), 101–118.

Schreiber, C. (2014). *Curricula and the making of the citizens. Trajectories from 19th and 20th century Luxembourg.* (PhD Thesis), University of Luxembourg, Luxembourg. (PhD-FLSHASE-2014-17)

Skovsmose, O. (1994). *Towards a philosophy of critical mathematics education.* Dordrecht, the Netherlands: Kluwer.

Skovsmose, O., & Valero, P. (2001). Breaking political neutrality: The critical engagement of mathematics education with democracy. In B. Atweh, H. Forgasz, & B. Nebres (Eds.), *Sociocultural research on mathematics education. An international perspective.* (pp. 37–55). Mahwah, NJ: Erlbaum.

Skovsmose, O., Valero, P. (2008). Democratic access to powerful mathematical ideas. In L.D. English (Ed.), *Handbook of international research in mathematics education. Directions for the 21st Century, (2nd ed.)* (pp. 415–438). Mahwah, NJ: Erlbaum.

Sriraman, B., & English, L. (2010). Surveying theories and philosophies of mathematics education. In B. Sriraman & L. English (Eds.), *Theories of mathematics education: Seeking new frontiers* (pp. 7–32). Heidelberg, Germany: Springer.

Stech, S. (2008). School mathematics as a developmental activity. In A. Watson & P. Winbourne (Eds.), *New directions for situated cognition in mathematics education* (pp. 13–30). New York, NY: Springer.

Valero, P. (2004). Socio-political perspectives on mathematics education. In P. Valero & R. Zevenbergen (Eds.), *Researching the socio-political dimensions of mathematics education: Issues of power in theory and methodology* (pp. 5–24). Boston, MA: Kluwer.

Valero, P. (2014). Political perspectives in mathematics education. In S. Lerman (Ed.), *Encyclopedia of mathematics education* (pp. 484–487). New York, NY: SpringerReference.

Valero, P. & Stentoft, D. (2010). The 'post' move of critical mathematics education. In H. Alrø, O. Ravn & P. Valero (Ed.), *Critical mathematics education: Past, present and future* (pp. 183–196). Rotterdam, the Netherlands: Sense.

Valero, P., García, G., Camelo, F., Mancera, G., & Romero, J. (2012). Mathematics education and the dignity of being. *Pythagoras, 33*(2). Retrieved from http://dx.doi.org/10.4102/pythagoras.v33i2.171

Vinner, S. (1997). From intuition to inhibition—mathematics education and other endangered species. In E. Pehkonen (Ed.), *Proceedings of the 21th conference of the International Group for Psychology of Mathematics Education*, (Vol. 1) (pp. 63–78). Helsinki, Finland: Lahti Research and Training Centre, University of Helsinki.

Walkerdine, V. (1988). *The mastery of reason: Cognitive development and the production of rationality.* London, England: Routledge.

Walshaw, M. (Ed.) (2010). *Unpacking pedagogies. New perspectives for mathematics.* Charlotte, NC: Information Age.

Žižek, S. (1989/2008). *The sublime object of ideology.* London, England: Verso.

Paola Valero
Aalborg University
Denmark

Alexander Pais
Manchester Metropolitan University
United Kingdom

CHAPTER 11

MATHEMATICAL ANALYSIS IN HIGH SCHOOL

A Fundamental Dilemma

Carl Winsløw

INTRODUCTION

Analysis, as construed here, is a domain of mathematics that treats problems related to limits, real and complex functions, and linear operators. While some of these problems have been known for thousands of years, the fundamentals of contemporary analysis—which includes a rigorous theory of real numbers—have been established over the past 400 years. Analysis has been closely linked to geometry and algebra, and also to a number of domains in the natural and social sciences. In particular, theoretical constructs like derivative and integral, have been historically linked to fundamental notions in mechanics and geometry (such as speed and area); while today derivatives and integrals are used in many other contexts.

The introduction of elementary analysis into secondary-school level mathematics, especially differential and integral calculus, has been historically justified by the manifest and increasing importance of these elements, both in pure mathematics and in other disciplines. How and when it is

Refractions of Mathematics Education, pages 197–213

done has clearly varied from one national or regional context to another; for instance it remains an option in the United States (Spresser, 1979). In Denmark, the first timid introduction of infinitely small and large quantities as a mandatory topic in high school, came as early as 1906. The teaching of infinitesimal calculus—that is, differential and integral calculus—has been mandatory in the scientific stream since 1935. Ever since then, the investigation of functions based on derivatives and integrals has remained a relatively stable and central part of the tasks posed for the national written examinations in the scientific streams of Danish high school (Petersen and Vagner, 2003). And it certainly remains a central element of the more advanced mathematics curriculum at this level. Most of the analysis exercises from the national final exams of the late 1930s, could be found in today's exams, except for details of formulation.

Despite the stability of the core types of tasks—such as determining the extreme values of a given function on an interval—one should nevertheless point out two major periods of change which are not specific to Danish high schools, but can be found in many European countries:

- Around 1960, the range and formality of mathematical themes was significantly extended, especially in adjacent domains such as set theory and algebra; but all or most of the extensions were subsequently abandoned after a decade or two.
- From around 1980, the progressive introduction of calculating devices in secondary schools, has increasingly affected the teaching of certain core techniques in analysis.

In this chapter, we will first provide a theoretical framework for analyzing and comparing different forms of organizing introductions to mathematical analysis; then we will illustrate it by two characteristic examples from the above periods of change as they occurred in Denmark, which will be based on the national exam tasks and textbooks used in the two periods. We will conclude by extracting from this a fundamental dilemma for the teaching of analysis at the secondary level, in view of the requirements and availability of computer algebra systems on the one hand, and contemporary utilitarian school pedagogies on the other hand.

AN EPISTEMOLOGICAL REFERENCE MODEL

As affirmed already in the first phrase of this chapter, the notion of limit is fundamental to mathematical analysis; and in particular to elements that appear in most introductory calculus teaching. Among these elements, the most important is probably the notion of derivative function; and in order to present an explicit general definition of derivation, any calculus text will have to introduce at least an informal explanation of what $\lim_{x \to a} f(x)$ means for a function f defined in a neighborhood of a (except possibly at a). Of course, the actual computation of such limits may also be useful for investigations of the function f itself, as well as in other contexts. As a result, most calculus texts and syllabi include at least a little practice and theory related to limits, prior to the introduction of derivatives.

For their study of the teaching of limits of functions in Spanish high school, Barbé, Bosch, Espinoza and Gascón (2005) proposed an epistemological reference model; this would trace the didactic transposition of pertinent knowledge, whose end result was the didactic process observed in the classroom. As this study has appeared in a widely accessible journal, we will recall only one main point; while we use the full theoretical framework explained in that paper, in particular the basic notions of the anthropological theory of the didactical presented in section 2 of the paper.

The main point we wish to emphasize, is that the study's authors identified two local mathematical organizations, which they used for their analysis of the different stages of the didactic transposition of the basic theory of limits in a Spanish high school class. These organizations were:

- MO_1, termed "the algebra of limits," where the practice is unified by a discourse about how to compute limits in a variety of cases, and the practice blocks amount to such cases: each consists of a type of task with a technique that allows one to solve all tasks of the given type—for instance, to compute $\lim_{x \to a} f(a)$ when f is a polynomial, the answer is simply $f(a)$. At the theoretical level of this organization, there are algebraic rules like $\lim_{x \to a}(f(x) + g(x)) = \lim_{x \to a} f(x) + \lim_{x \to a} g(x)$ which are not further justified. Also the existence of limits is not problematized beyond the possibility of computation.
- MO_2, termed "the topology of limits," where the practice is unified by an abstract discourse and theory about limits, including for example a rigorous definition of what it means for $\lim_{x \to a} f(x)$ to exist; the types of tasks in this organization include determining if a given function has a limit at a given point, and to justify calculation rules like $\lim_{x \to a}(f(x) + g(x)) = \lim_{x \to a} f(x) + \lim_{x \to a} g(x)$ under appropriate assumptions.

The link between the two organizations is clear, at least in principle: the practice block of MO_2 is needed to justify the theoretical level of MO_1 in a wider theoretical context (while, locally, the calculations rules for limits might be regarded as a kind of self-evident axioms). In this wider theoretical sense of limits, namely that of academic mathematics, one might even say that MO_2 comes first: before calculating $\lim_{x \to a} f(x)$ we need to define what it means, and that certainly includes non-trivial conditions for existence.

In didactic practice MO_2 does not need to come first. It is apparent from the study cited above, as well as from other research on the teaching of limits (often with less explicit reference models) that the practice block of MO_1 may in fact be taught and learned with relative ease and efficiency. This can be done by using a theoretical block that is limited to informal and practice oriented explanation of the calculation rules. On the other hand, the teaching of MO_2 is usually absent or sparse at the secondary level, both in Spain and elsewhwere. In fact, convergence is often described informally, based on examples of function graphs and verbal explanations of how the function value gets close to a limit value as the free variable moves towards a given value. The development of a practice block (with mathematical techniques related to MO_2) for students is quite rare; it would, for instance, imply giving students rigorous techniques to decide on the question of existence of $\lim_{x \to a} f(x)$ in concrete and non-trivial cases. By rigorous, we mean that the technique can be explained and justified at the theory level of MO_2, such as the example given in Barbé et al. (2005: p. 243). When it is done, it is often prepared by introducing first the simpler theory of *limits of sequences* of real numbers.

We note here the strange and almost circular use of the term *continuity*, which is found in the Spanish high school (Barbé et al., p. 255) and most likely in many similar institutions. The meaningfulness of this notion seems to be particularly affected by the lack of a practical block in the didactic transposition of MO_2, both in the prescribed and realized mathematical organization.

Our epistemological reference model is based on the contention that a similar divide can be described and observed concerning other key elements of secondary level calculus, namely derived functions and integrals. In fact, when we consider the following (rough) definition

$$f'(x) = \lim_{h \to 0} \frac{f(x+h) - f(x)}{h}$$

we see immediately that the definition of derivatives and the justification of the rules governing their behavior, may indeed be considered as generating a local mathematical organization, which is directly derived from MO_2 as described above. More generally, the definition of the derivative generates a technology unifying a local mathematical organization MO_4 whose most basic types of task are, for a given function; to describe what $f'(x)$ is, to determine whether it exists, and to justify the so-called "rules of differentiation." These rules also constitute the theory level of an algebra of differentiation MO_3, which (as before) can exist in relative independence from MO_4.

We should not fail to note here that important theoretical results in differential calculus—like the mean value theorem—rely not just on MO_2, but also on other local organizations unified by a theory on the real number system. And some of these results are indeed important to justify other basic elements of secondary level analysis (like the link between the derivative of f and the monotonicity of f). So even for the purpose of analyzing secondary-level analysis, an epistemological reference model could not consider the theory of derivatives as merely derived from MO_2.

Another significant difference—not least for didactic transpositions—is that ultimately differentiation is an operation that (from a given function) produces another function; not just a number, as in the case of limits of function at a point. This need to think of functions as objects is further accentuated in the case of differential equations, and it has been extensively discussed in the literature on presumed cognitive obstacles to calculus (Tall, 1996). It seems plausible that it is also a didactic obstacle, because the mathematical organizations encountered by students before MO_3, usually do not have practical blocks with functions as algebraic objects (i.e., objects to be calculated with, and that are legitimate as answers).

While an exhaustive model is not the main aim here, we contend that other local organizations of differential calculus (such as those based on optimization tasks or to the solution of differential equations) can also be described in terms of an algebraic local organization (related to computational tasks), and a topological one (related to the definitions, conditions and justifications of what and how the computation is done).

Finally, the last "grand object" of secondary-level analysis, is the definite integral. Again there are two basic questions to be asked, given a function defined in an interval $[a, b]$: does the integral

$$\int_a^b f(x)\,dx$$

exist, and if so, how do we find it? From the academic mathematics point of view, this is related to what Jablonka and Klisinska (2012) investigated as the meaning of the fundamental theorem of calculus, in history as well as in the minds of contemporary mathematicians. With several possible variations in the formulation, this theorem provide answers to the two basic questions just mentioned, and states that:

1. If f is continuous on I, then f has an antiderivative on I; and if f has an antiderivative on I, then f is integrable on I.
2. If F is an antiderivative to f on I, then

$$\int_a^b f(x)\,dx = F(b) - F(a).$$

The said variations in the formulation of the theorem are less interesting for its meaning than how one defines

$$\int_a^b f(x)\,dx$$

to begin with. In fact, many textbooks (both for secondary and tertiary level) use part (2) of the theorem as a definition (i.e., they define the integral in terms of an antiderivative). Then, of course, the theorem reduces to the first half of (1), which becomes merely a claim. Still, one has an excellent new local organization MO_5, the algebra of integration, with rules that are, even at the theoretical level, easily justified from the rules at the theoretical level of MO_3. This also suffices for the needs of some of the more advanced local organizations of differential calculus, like the algebra of solving separable differential equations. In fact, this definition works well as long as one does not seek any separate meaning in the number

$$\int_a^b f(x)\,dx$$

—or in the fundamental theorem of calculus.

Of course most introductions of the integral also relate it to area. And in some contemporary textbooks, one finds a slightly different approach to defining the integral: for a positive function f it is defined as the *area* of the point set $\{(x,y): a < x < b, 0 < y < f(x)\}$ while assuming tacitly or informally that this area makes sense for "good functions." Clearly, this is just like defining the limit informally: the definition makes sense in an intuitive way, but it does not suffice to enable a mathematical practice block related to MO_6, such as deciding on the existence of the object defined, or justifying the basic rules and properties satisfied by this area integral. This entrance to integrals does not need to leave the link to the integral by derivatives entirely in the dark: if one accepts the definition of the area integral above as meaningful in itself, one may show from the definition of derivatives that the area—when viewed as a function of b—is an anti-derivative of the function f. This is indeed done in many contemporary Danish textbooks for upper secondary school, and it has undoubtedly been repeated by students thousands of times at the oral part of the mathematics exam.

Even in academic (or scholarly) mathematics, the integral is defined in different ways; and development of alternative approaches, offers an interesting chapter in the history of analysis, as exposed by Jablonka and Klisinska. While Lebesgue integration is often considered superior for advanced purposes, a more basic approach is the one due to B. Riemann, and a didactic transposition of it to high school will be explained in the next section.

But with any rigorous definition of integrals, the topological counterpart to MO_5 appears on the scene: a local organization MO_6 unified by the theoretical definition of the integral, linked to the fundamental properties of the real number space, and with the practical block being concerned with the tasks of deciding on the existence of the integral, and with justifying the rules governing its calculus.

With this, we have extended the epistemological reference model from Barbé et al. to cover the elements of secondary-level analysis (or, in some countries, the introduction to university level calculus); the result is illustrated in Table 11.1. The point is that introductory analysis can be roughly modeled as pairs of local mathematical organizations—algebraic and topological ones—which team up in regional ones, and build on each other more or less in the sequence shown. The algebraic organizations, exhibit practical blocks with algorithmic techniques that can be taught and learned if not with ease, then at least in an orderly fashion, task type by task type. It is this part that is called calculus in American textbooks, corresponding to certain practice blocks of introductory real analysis. On the other hand, the meaning of it all is related to topological definitions and properties, which are also needed for a deeper justification of the calculus, but which are less evident to transpose to the classroom, because of the ultimate reliance on a complete theory of the real numbers.

We have already pointed out that the six local organizations presented above and in Figure 11.1, do not exhaust even the most modest version of analysis at the secondary level. Also the task types in the table, are declined into smaller collections of tasks in reality, each characterized by one

TABLE 11.1   A Model for Secondary-Level Analysis: Local Organizations and Basic Task Types

Object	Existence/"topology"	Computation/"algebra"
Limit of function f at point $a \in [-\infty, \infty]$	MO_2 T_{21}: Does $\lim_{x \to a} f(x)$ exist? T_{22}: Justify rules and properties $\rightarrow$	MO_1 T_{11}: Find $\lim_{x \to a} f(x)$. THEORY BLOCK
Derivative of function f	MO_4 T_{41}: Does f' exist? Where? T_{42}: Justify rules and properties $\rightarrow$	MO_3 T_{32}: Find f'. THEORY BLOCK
Integral of function f on interval $[a, b] \subseteq [-\infty, \infty]$	MO_6 T_{61}: Does $\int_a^b f(x)\,dx$ exist? T_{62}: Justify rules and properties $\rightarrow$	MO_5 T_{51}: Find $\int_a^b f(x)\,dx$. THEORY BLOCK

technique. So the role of the model presented here is not to be comprehensive or give all details; but instead to help us articulate principal and crucial challenges for any didactic transposition of analysis, and in particular to support our reflections on the meaning and character of the two recent "major changes" mentioned in the introduction.

THE CASE OF INTEGRATION : A DIDACTIC TRANSPOSITION FROM THE PAST

The most eye-catching changes of the 1961 reform of Danish high school, was the introduction of elements of logics, set theory and abstract algebra. Some of these elements can be made useful also to define and study functions in the domain of analysis. As was mentioned in the introduction, the reform did not dramatically affect the tasks related to analysis, which appear in the final written examinations of Danish high school; although an increase in variation and difficulty of exam tasks has been evident. The novelties in abstract algebra are more visible, even in the analysis tasks (with a siginificant change in terminology from curves to functions as the objects to be examined). In terms of our epistemological reference model, the exam tasks all relate to the practice blocks of the algebraic organizations MO_1, MO_3 and MO_5. A typical exam exercise is the following from 1971 (Petersen & Wagner, 2003, p. 256):

A function f is given by $f(x) = xe^{-2x}$, $x \in \mathbb{R}$, where $\mathbb{R}$ designates the set of real numbers.

Investigate f as regards its zeros, sign and monotonicity.

Determine the area of the point set given by $\{(x,y) | 0 \leq x \leq \frac{1}{2} \wedge 0 \leq y \leq f(x)\}$

For any positive real number a a function g_a is given by $g_a(x) = xe^{-ax}$, $x \in \mathbb{R}$.

Show that g_a has a maximal value and find it.

As in many other tasks, the analysis appears in the investigation of certain properties of a given function; for instance, to find its asymptotes (reduces to find one or more limits, i.e., to MO_1-tasks); to determine monotonicity or suprema (the key to which is to find f', i.e., an MO_3-task); or to determine the area or volume of certain figures (reduces to a definite integral, i.e., an MO_5-task). All of these tasks continue to be common at the written examinations.

Textbooks from the period 1961–1980 reveal more profound additions to the theory blocks taught, and certainly also marked differences with contemporary teaching at this level. In fact, all six local organizations described above are covered in detail, both in exposition of theory, in worked examples, and in exercises. Today's university students of mathematics usually refuse to believe that this can be done at the secondary level, because it is now part of their first year. To show that, and how it is really done, I will provide a rather extensive exposition of the presentation of integration in a textbook series authored by Kristensen and Rindung (1973), which dominated Danish high schools from the late 1960s to the early 1980s. Here we will study the first of the two books written for the second year of high school, and only the second edition from 1973. This edition differs from the 1963 edition in several respects; most notably it has a less rigorous treatment of the topology of the real numbers. For instance, in the 1973 edition, all mention of *supremum* and *infimum* had been dropped. This clearly affected also the introduction of the definite integral (in fact, the Riemann integral), which we now present.

The chapter on integration had 12 main sections (we provide a short description in parentheses):

- **Area** (8 pages): An informal discussion of area of non-polygonal point sets, and how it may be approached through double approximation with polygonal point sets.
- **Mean sums, upper sums, lower sums** (5 pages): A rigorous definition of these notions for bounded functions on an interval, ending with the theorem that every lower sum is less than every upper sum.
- **Integrability** (3 pages): Rigorous definition by *the existence of a unique number situated between all lower sums and all upper sums*; proof that every monotonous function is integrable.
- **The integral and mean sums** (3 pages): Proof that the integral, if it exists, is a limit of mean sums and can be considered as a "mean value" of the function on the interval.
- **Interval additivity theorem** (4 pages): With a proof based on the above definition.
- **The class of integrable functions** (3 pages): Applies additivity theorem to prove that piecewise monotonous functions are integrable. A discussion of examples and more general results, including the theorem that continuous functions are integrable—stated without proof.
- **Integral and antiderivative** (4 pages): With proof that if a function is integrable and has an antiderivative, then the formula above applies.
- **Existence of antiderivatives** (2 pages): Proof that if a function f is continuous on an interval I and $a \in I$ then

$$F(x) = \int_a^x f(x)\,dx$$

is an antiderivative to f on I.

- **The indefinite integral** (1 page): Introduction of the symbol $\int f(x)\,dx$ for the class of antiderivatives.
- **Calculation rules for integration** (15 pages): Including substitution and parts, with many examples.
- **Existence of logarithm functions** (2 pages): Continuation of a "gap" left in the first year volume, filling it by proving that the integral of $1/x$ gives a function with the previously stated properties.
- **Application of integral calculus** (10 pages): Including volumes, curve length and examples from physics and financial theory.

As this outline shows, the text presents techniques and theory covering most of MO_6, with the single exception that integrability is only shown for piecewise monotonous functions, not for general continuous functions. It essentially presents this *before* MO_5, and approximately the same space is allowed for each of these local organizations, the main link being the justification that integrals of common functions can be computed by antiderivatives. For practical purposes, integrability is certainly sufficiently covered, as all functions normally considered in high school are piecewise monotonous, even if the book does present one continuous function that is not. The need to state the theorem about integrability of continuous functions, is due to its use to prove the existence of antiderivatives. This is an important point in view of MO_5, since there are simple and common functions to which none of the "calculations rules" succeeds in producing the antiderivative; unlike differentiation and limits, it thus appears harder to dismiss the existence problem with the notion that, "we only consider functions where the algebraic rules apply."

Clearly, the textbook exposition of theory from MO_6 does not in itself guarantee that students will engage in any related practice, besides absorbing and reciting proofs on demand. So a really interesting feature of the chapters dealing with the topology of integrals is the attempt to engage students in solving tasks. Here are some examples of exercises from this same textbook series:

301. Show that f given by

$$f(x) = \begin{cases} x, x & \text{rational} \\ 2x, x & \text{irrational} \end{cases}$$

is not integrable on $[0,1]$.

308. Assume that f is a bounded and integrable function on the interval $[a, b]$. The function F is defined by

$$F(x) = \int_a^x f(t)\,dt$$

for any x in $[a, b]$. Show that F is continuous.

Indeed, to solve the first exercise, students must show that every lower sum is smaller than one half, while every upper sum is at least 1; that is, they will mobilize a genuine MO_6-technique (to show non-integrability based on the definition mentioned above). Similarly, the second exercise requires putting the interval additivity theorem to use, together with techniques related to inequalities (a central part of MO_6). It is indeed possible that both the practice and theory related to MO_6 had not been studied with the same intensity by all classes at the time. In fact, the national written exam concerned exclusively MO_5; in particular, not one exam task even asked for proving the integrability of a function. And the final oral exam was more concerned with theorems and proofs; the teacher always had some freedom to select emphases and topics. However past syllabi (Petersen & Vagner, 2003, p. 243), as well as the author's personal memory, confirm that both the theory and practice of MO_6 were certainly developed according to the ambitions of the textbook and its task inventory. But as noted by experienced teachers (Petersen & Vagner, 2003), over the 1970s the students increasingly had difficulty appreciating the cautious and stringent fashion in which topics were treated in Kristensen and Rindung. The reason was, among other things, the learning-by-doing pedagogy that grew in importance in primary and lower secondary school.

It is evident that the new-math period ended in a more quick and abrupt way in Danish primary and lower secondary school. In high school, the use of Kristensen and Rindung continued well into the 1980s; the author of this chapter remembers working the two exercises quoted above in 1984. This difference is not unrelated to the fact that Danish primary and lower secondary level teachers do not study mathematics at universities, and as a result, they have little or no experience with modern mathematics. But no doubt is was also important that there are different external constraints on the two types of school institutions.

THE CASE OF INTEGRATION: AN EXAMPLE OF A RECENT DIDACTIC TRANSPOSITION

After the major reforms of the 1980s and 1990s, Danish high school has become more diversified with several streams and options; this makes it more difficult to describe a typical approach to a sector like integration. The general tendency, already alluded to above, is clearly that MO_6 is not taught, except for mentioning the link between the definite integral and certain "areas" which are assumed to make sense as a piece of nature. Clearly, MO_5 has become even more dominant, but it has also changed, as the use of symbolic calculators for integration is now both allowed and taught, along with non-instrumented techniques that are still required in parts of the final written exam (see Drijvers, 2009, for a more detailed study of CAS-use in final high-school exams in the Danish and other contexts). However, it may still be possible to study informal techniques related to MO_6 as so-called optional topics. We have chosen to present some ideas from a textbook by Bregendal, Schmidt and Vestergaard (2007), which illustrates how this can be done in continuation of the mandatory material.

The book has two chapters on integration: the first covers the mandatory material and the second deals with more advanced options. The latter includes, nowadays, MO_5 techniques like integration by parts; but the chapter also features a 10-page section on numeric integration, which is the excerpt we will consider here; it is the part of the book that comes closest to asking the question about the existence of integrals.

The section in question opens with an informal description of how Archimedes approximated π, the area of the unit circle, by computing the areas of inscribed and circumscribed regular polygons.

The authors go on to explain how left and right sums, which correspond to the areas of certain rectangles, can be used to do something similar in order to compute the area of $\{(x, y) : 0 < x < 1, 0 < y < x^2\}$, and that the average of these two sums (corresponding to trapezes through the middle points of the rectangles) gives a good approximation already for just four intervals. The good value ($\frac{1}{3}$) is known, because it has presumably been established that the area can be computed using the antiderivative. The authors then explain in great detail that the nth right sums are

$$H_n = \sum_{i=1}^{n} \left(\frac{i}{n} \right)^2 \frac{1}{n} = \frac{2n^3 + 3n^2 + n}{6n^3},$$

and a similar formula for the left sums is given (to be proved in an exercise). Both converge to $\frac{1}{3}$. They then state (p. 80)—with no justification—that:

A similar relation between these sums and the integral

$$\int_a^b f(x)\, dx$$

is valid for any continuous function f on an interval $[a, b]$, and we can in particular conclude that for such a function, one has

$$H_n = \sum_{i=1}^{n} f(x_i) \cdot \Delta x \to \int_a^b f(x)\, dx \text{ as } n \to \infty \text{ (that is, when } \Delta x \to 0).$$

The authors proceed to show a graph of a non-specific concave function (see Figure 11.1), and point out that the "comparative size of the left and right sums depends on the form of the graph." An attentive reading of the

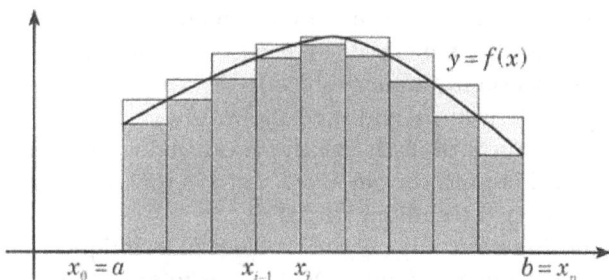

Figure 11.1 Illustration of right and left sums (Bregendahl et al., 2007).

figure shows that neither is clearly smaller or greater than the integral. This blurs the connection to the idea of Archimedes. Still, an informal connection between the limit of a sum and the area, has been established for a concrete increasing function, where right and left sums do enclose the area to be computed.

At the end of the section, some concrete worked examples and exercises are given about how to compute right and left sums using the statistics package of a calculator (Texas TI-84) for large values of n. No use of graphical visualization is suggested at this point.

In terms of our reference model, one could first think that the authors really sought to give an informal treatment of the Riemann integral; this impression was confirmed by a historical note in the book margin. It has a picture of B. Riemann and a text claiming, among other things, that the German Riemann clarified the properties of a function that make it integrable. For this reason, the integral we have worked with is also called the Riemann integral.

However, nowhere else was the notion of integrable mentioned in the text. It was never said that the existence of the limit was a condition for the integral to exist; it was merely postulated that the formula was "valid" (quote given above). We did not get near to a distinction between lower, upper, middle, right, and left sums; this would certainly be needed for a more formal treatment of the Riemann integral. Only the two last sums were dealt with; but they did not really correspond to the idea of Archimedes, which introduced the section, or to the main idea behind Riemann's integral.

In fact, the main point seems to have been to give an alternative (and potentially instrumented) technique for computing the integral, namely the informally topological formula for the integral (as the limit of sums). While the algebraic techniques used to compute H_n for the case $f(x) = x^2$ on $[0, 1]$ will certainly not go much beyond that example, the numeric technique establishes a kind of experimental relation between a limit process (infinite sum) and the integral defined (and computed) using antiderivatives. The book noted some pages earlier that it was sometimes impossible to find an antiderivative, "using the methods we have seen," and that this then motivated the introduction of "numerical methods." As the integral is originally *defined* using antiderivatives, and then shown to correspond to an area in some cases, the limit formula is in fact introduced as an alternative technique for *finding* integrals (i.e., in MO_5), not as a tool to *define* integrability and integrals (i.e., as a technique for MO_6).

In short, the epistemic value of the topological technique, which is at the root of Riemann's integral (and of MO_6), fails to appear in the text. And the pragmatic value of the alternative numerical technique (whether instrumented or not), may be equally unconvincing to students in possession of calculators who can do numerical computations of definite integrals in one step. Of course, the same can be said of most of the techniques of MO_5; any symbolic integration that manual techniques can achieve, is also done in one step by the students' CAS devices.

THE DILEMMA —AND A CHALLENGE

Calculators became mandatory tools in Danish high school mathematics around 1980. At first, these handheld devices replaced tables and other tools for computing values of special functions etc., thus suppressing previously important techniques and tools. During the 1980s, a rapid succession of more advanced calculators appeared: programmable, graphical, and computer-algebra-system (CAS) calculators. With more or less delay, the use of some or all of these devices—as well as similar laptop software—has become part of the high-school teaching of mathematics. This is not the place to go into the historical or didactical subtleties of this development (we refer to Hoyles and Lagrange, 2010). We just stress that the CAS systems that are used now in high school teaching, at least in Denmark, provide instrumented techniques of high pragmatic value (or efficiency) for all basic tasks in MO_1, MO_3 and MO_5. Most standard tasks related to investigating given function are simplified, if not trivialized, using these techniques. At the same time, the connection between local organizations—such as the key connection between MO_3 and MO_5 as opposite tasks—tend to disappear when considering these organizations with instrumented techniques. It is possible and necessary to develop new tasks for students, both for the daily teaching and final examinations. Indeed the interpretation of more or (often) less authentic situations in terms of function models, is a much treasured direction for doing so, at least in Denmark (Drijvers, 2009). But it cannot be denied that there is a certain dissatisfaction in terms of the mathematical content. When the computation of limits, derivatives, maximal and minimal values, antiderivatives, and so on, is reduced to independent, one-key operations, not much is left of the algebraic organizations and their theoretical coherence. At the same time, the topological organizations MO_2, MO_4 and MO_6 have already been long abandoned, at least in the formally demanding transpositions they used to have.

The dilemma we then face is the following: what used to be the core of high school mathematics for decades (almost 80 years in Denmark), seems now to have been reduced to a collection of independent, highly instrumented techniques, together with a basic algebraic technology of functions and numbers. This then has enabled them to be used and combined to solve a variety of variation problems, which are important in many settings, including extra-mathematical ones. The theory of computation (corresponding to rules of the theory blocks of MO_1, MO_3, and MO_5) continues to be taught and learned in a more or less complete and abstract form, but its practical value is to a large extent gone, at least for beginners. We have already explained the incoherence resulting from the elimination of the topological parts of mathematical analysis in the transposition to high school mathematics; with the introduction of instrumented techniques, we may face a more or less complete collapse, when it comes to the coherence that remains among (and inside) the local algebraic organizations.

Of course the dilemma can also be considered as a challenge: how can we reorganize or modernize the transposition of mathematical analysis to and in high school teaching, in ways that make use of the technology, while presenting the mathematical domain of analysis in a more complete and satisfactory way (than as a set of modeling tools)?

The answer of the past transposition by Kristensen and Rindung (1973), was essentially to keep as closely as possible to the scientific mathematics of its time. Clearly, the pedagogical and political trends have made that principle less evident today. But it should at least be noted that the proximity principle of the past might not lead to exactly the same answers as it did in the 1960s. One reason is that CAS-based experimental methods have also become part of the scientific practice to the scientist, who develops and uses mathematical analysis. The heart of analysis, which remains with the limits and deep properties of the real number systems, can be accessed and treated in new ways using technology; this can begin with somewhat ostensive approaches (for instance, as applets "showing" definitions, e.g., www.maplesoft.com/products/mapleplayer/). Another reason is that the mathematical analyst of today is yet another generation from the time when rigorous analysis was something new and exciting: functions, limits, and the other key objects have somehow been tamed by the mathematical practice, just like complex and negative numbers a little earlier. This could lead to a higher tolerance for relative informal approaches in the secondary curriculum, as long as the transposition preserves what the contemporary scientist regards as essential to the transposed mathematical organizations.

The answer for the recent transposition of the Riemann integral, which was presented previously, appears clearly unsatisfactory, even if it hints at the potential interest of a sequence approach to the topological side of elementary analysis. The dilemma was only accentuated by adding another more or less unjustified technique to the transposition of MO_5. An interesting alternative would be to introduce the integral *first* as the limit of (say) right sums, with convergence being the condition of existence; *then* derive some of the properties that allow for (some sort of) proof that the derivative of the integral is the integrand. A related but much more radical alternative would be to revert the common transposition and present integration (including *both* MO_5 and MO_6) before differentiation. This approach has already been completely developed by Apostol (1967), who advocated the choice by appealing to the historic precedence of problems related to integration (in fact, the Archimedean arguments alluded to previously). I do not know of any high-school textbooks taking this approach, which still seems to appear almost offensive to some college teachers (Math Forum, 2009). However, the use of instrumented techniques for computation and visualization, and the many attractive uses of integration, could well mean that teaching this sector first, might become an interesting option at the secondary level.

By way of conclusion, the teaching of analysis in secondary school is not only threatened in its time-honored form by the availability of new technology. In fact, the exercise of certain algebraic techniques, as the main element in students' praxeologies in analysis, had already become critically separated from the mathematical and extra-mathematical questions that motivate their development; and also from each other, in the absence of theoretical elements that could help to relate and justify them as mathematical practices. Research and development concerning the secondary curriculum in analysis should not only focus on the algebraic side (despite the obvious interest of technology in this setting), but it should also seek ways in which mathematical software and other resources can help rebalance the fundamental synergy between algebra and topology in this topic. This means, in particular, to give students access to its fundamental constructs—limits, derivatives and integrals—in ways that will make them useful tools to solve problems that involve infinite sums, mean values, growth rates, and more.

REFERENCES

Apostol, T. (1967). *Calculus 1*, (2nd ed.). New York, NY: John Wiley.

Barbé, J., Bosch, M., Espinoza, L., & Gascón, J. (2005). Didactic restrictions on teachers' practice—The case of limits of functions in Spanish high schools. *Educational Studies in Mathematics, 59,* 235–268.

Bregendahl, P., Schmidt, S., & Vestergaard, L. (2007). *Mat A hhx.* Aarhus, Denmark: Systime.

Drijvers, P. (2009). Tools and tests: technology in national final mathematics examinations. In C. Winslow (Ed.), *Nordic research on mathematics education, Proceedings from NORMA08* (pp. 225–236). Rotterdam, the Netherlands: Sense.

Hoyles, C., & Lagrange, J.-B. (Eds.) (2010). *Mathematics education and technology-rethinking the terrain. The 13th ICMI study*. New York, NY: Springer.

Jablonka, E., & Klisinska, A. (2012), What was and is the fundamental theorem of calculus, really? In B. Sriraman (Ed.), *The Montana Mathematical Enthusiast Monographs No. 12: Crossroads in the history of mathematics and mathematics education* (pp. 3–40). Charlotte, NC & Missoula, MT: Information Age & Montana Council of Teachers of Mathematics.

Kristensen, E., & Rindung, O. (1973). *Matematik 2.1*. Copenhagen, Denmark: G. E. C. Gad.

Math Forum @ Drexel. (2009). Topic: Integration before differentiation. Accessed from http://mathforum.org/kb/message.jspa?messageID=6858531

Petersen, P., & Vagner, S. (2003). *Studentereksamensopgaver i matematik 1806–1991* [Final exam tasks in high school mathematics]. Hillerød, Denmark: Matematiklærerforeningen.

Spresser, D. (1979). Placement of the first calculus course. *International Journal for Mathematics in Science and Technology, 10*, 593–600.

Tall, D. (1996): Functions and calculus. In A. J. Bishop et al. (Eds.), *International handbook of mathematics education* (pp. 289–325). Dordrecht, the Netherlands: Kluwer.

Carl Winsløw
Copenhagen University
Denmark

CHAPTER 12

THE WHITE PAPER

ABSTRACT

Refractions of Mathematics Education, pages 215–223
Copyright © 2015 by Information Age Publishing
All rights of reproduction in any form reserved.

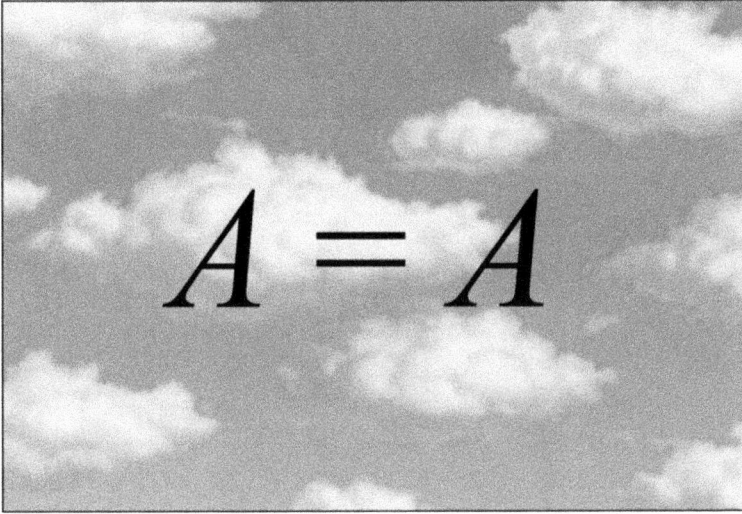

Figure 12.1a   This is not an equation.

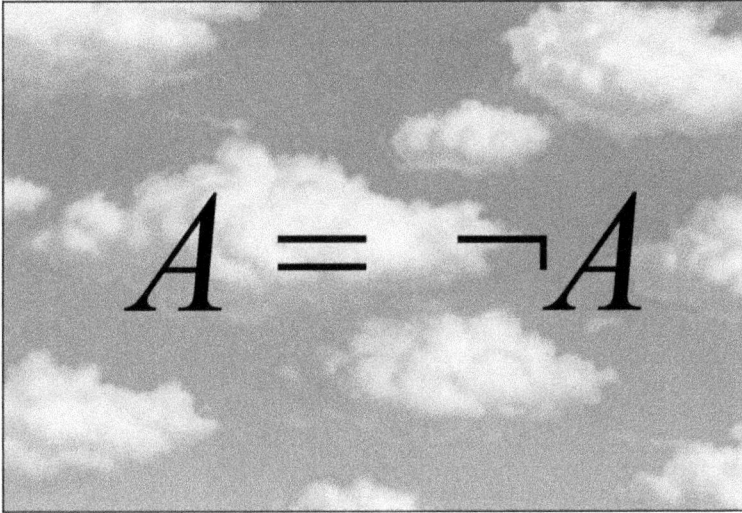

Figure 12.1b   This is an equation.

APPENDIX 12.1   QUESTIONNAIRE

	Yes	No	Don't Know
I am learning to obtain a position of power.	O	O	O
I am learning to help me attract others fitting my sexual preferences.	O	O	O
I am learning to earn more money.	O	O	O
I am learning because I am being paid for it.	O	O	O
I am learning because else I'd have to find myself a job.	O	O	O
I am learning because the law says I must go to school.	O	O	O
I am learning because I have been told to do so.	O	O	O
I am learning because I have not learned to do anything else.	O	O	O
I am learning because I have nothing else to do.	O	O	O
I am not learning anything.	O	O	O
Am I learning anything?	O	O	O
Don't know!	O	O	O

Other:

APPENDIX 12.2   GLOSSARY

abhava non-being
adiaphoros indifferent
adoxastos without opinions
akataplexia intrepidness, not able to grasp
aklineis firmly balanced, without agitation
akradantos without agitation, firmly balanced
anabhilapya inexpressible
anatta not-self, non-essence
anekantavada indeterminacy
anepicritos not able to judge
anicca impermanence
anirvacaniya undefinable
anitya impermanent, without self-nature
aoristia lack of boundary or definition
apatheia non reactiveness
aphasia non involvement in linguistic category projection
aponia non pain
aprapancita indiscriminate
arithmasthenia getting it wrong
arithmetic counting backwards
arrepsia lack of inclination
asthatmetos unstable
asunya non emptiness
ataraxia imperturbable
atarkavacara beyond logical argument
athambia inability to be astonished
autarkeia self-rule
avyakrta indeterminable

BIBLIOGRAPHY

Foucault, M. (1979). What is an author? In J. V. Harari (Ed.), *Textual strategies: Perspectives in post-structuralist criticism,* (pp.141–160). Ithaca, NY: Cornell University Press.

Freud, S. (1930). *Das Unbehagen in der Kultur* [Civilization and its discontents]. Vienna, Austria: Internationaler Psychoanalytischer Verlag.

Feyerabend, P. (1975/1988). *Against method* (Rev .ed.). London, England: Verso.

Haraway, D. (1988). Situated knowledge: The science question in feminism as a site of discourse on the privilege of partial perspective, *Feminist Studies 14* (3), 575–599.

Horkheimer, M., & Adorno, T. (1947/2002). G. Noerr (Ed.), E. Jephcott (Trans.). *Dialectic of enlightenment: Philosophical fragments.* Stanford, CA: Stanford University Press.

Jayarāśi Bhatta. (1940). S. Samghavī (Ed.), R. Pārīkh (Trans.). *Tattvôpaplava-simha* [The lion of the dissolution of (all) categories]. Baroda, India: Oriental Institute.

Loy, D. (1987). The clôture of deconstruction: A Mahāyāna critique of Derrida, *International Philosophical Quarterly 28*(1), 56–80.

McEvilley, T. (1982). Pyrrhonism and Mādhyamika, *Philosophy East and West 32,* 3–35.

Nussbaum, M. (1994). *Therapy of desire: Theory and practice in hellenistic ethics.* Princeton, NJ: Princeton University Press.

Sextus Empiricus. (1949/2000). R. G. Bury (Trans.). *Against the professors.* Cambridge, MA: Harvard University Press.

Sextus Empiricus. (1998). D. Blank (Trans.). *Against the grammarians (Adversos mathematicos I).* Oxford, England: Clarendon Press.

Sextus Empiricus. (2005). R. Bett (Trans.). *Against the logicians (Adversus mathematicos VII and VIII).* Cambridge, England: Cambridge University Press.

Sextus Empiricus. (2000). R. Bett (Trans.) *Against the ethicists (Adversus mathematicos XI).* Oxford, England: Clarendon Press.

Swift, J. (1726). *Travels into several remote nations of the world, by Lemuel Gulliver, first a surgeon, then a captain of ships. Part three: A voyage to Laputa, Balnibarbi, Luggnagg, Glubbdubdrib, and Japan.* London, England: Benjamin Motte.

The Academy of Lagado
Balnibarbi